HELP US TO HELP YOU!

We appreciate your use of this laboratory manual – one of an expanding range of technical publications from Cold Spring Harbor Laboratory Press.

In the future, we plan to announce new publications and services that will benefit all users of this manual. These include updates (in print and electronic form), new editions, and technical literature on related scientific topics.

To ensure that we can reach you with these announcements, and maximize the value of this lab manual, please take a moment to complete and return this registration card. Thank you for helping us to help you!

Name ______________________________

Address ______________________________

____

City/State/Zip ______________________________

Country ______________________________

FAX ______________ E-Mail ______________

Please check title(s) that most closely describe(s) your position:

____(1) Professor
____(2) Graduate student
____(3) Postdoctoral scientist
____(4) Lab director
____(5) Lab technician
____(6) Medical student
____(7) Undergraduate student
____(8) Librarian
____(9) Publisher

Please check your employment category:

____(1) University/college
____(2) Research institute/foundation
____(3) Hospital
____(4) Medical school
____(5) Industry
____(6) Government
____(7) Library/information center

Please check your primary field of interest:

____(1) Biochemistry
____(2) Cell biology
____(3) Developmental biology
____(4) Epidemiology
____(5) Genetics
____(6) Immunology
____(7) Microbiology
____(8) Molecular biology
____(9) Neurobiology
____(10) Plant biology
____(11) Pharmacology
____(12) Virology
____(13) Oncology
____(14) Other____________

HELP US TO HELP YOU!

We appreciate your use of this laboratory manual – one of an expanding range of technical publications from Cold Spring Harbor Laboratory Press.

In the future, we plan to announce new publications and services that will benefit all users of this manual. These include updates (in print and electronic form), new editions, and technical literature on related scientific topics.

To ensure that we can reach you with these announcements, and maximize the value of this lab manual, please take a moment to complete and return this registration card. Thank you for helping us to help you!

Name ______________________________

Address ______________________________

____

City/State/Zip ______________________________

Country ______________________________

FAX ______________ E-Mail ______________

Please check title(s) that most closely describe(s) your position:

____(1) Professor
____(2) Graduate student
____(3) Postdoctoral scientist
____(4) Lab director
____(5) Lab technician
____(6) Medical student
____(7) Undergraduate student
____(8) Librarian
____(9) Publisher

Please check your employment category:

____(1) University/college
____(2) Research institute/foundation
____(3) Hospital
____(4) Medical school
____(5) Industry
____(6) Government
____(7) Library/information center

Please check your primary field of interest:

____(1) Biochemistry
____(2) Cell biology
____(3) Developmental biology
____(4) Epidemiology
____(5) Genetics
____(6) Immunology
____(7) Microbiology
____(8) Molecular biology
____(9) Neurobiology
____(10) Plant biology
____(11) Pharmacology
____(12) Virology
____(13) Oncology
____(14) Other____________

HELP US TO HELP YOU!

We appreciate your use of this laboratory manual – one of an expanding range of technical publications from Cold Spring Harbor Laboratory Press.

In the future, we plan to announce new publications and services that will benefit all users of this manual. These include updates (in print and electronic form), new editions, and technical literature on related scientific topics.

To ensure that we can reach you with these announcements, and maximize the value of this lab manual, please take a moment to complete and return this registration card. Thank you for helping us to help you!

Name ______________________________

Address ______________________________

____

City/State/Zip ______________________________

Country ______________________________

FAX ______________ E-Mail ______________

Please check title(s) that most closely describe(s) your position:

____(1) Professor
____(2) Graduate student
____(3) Postdoctoral scientist
____(4) Lab director
____(5) Lab technician
____(6) Medical student
____(7) Undergraduate student
____(8) Librarian
____(9) Publisher

Please check your employment category:

____(1) University/college
____(2) Research institute/foundation
____(3) Hospital
____(4) Medical school
____(5) Industry
____(6) Government
____(7) Library/information center

Please check your primary field of interest:

____(1) Biochemistry
____(2) Cell biology
____(3) Developmental biology
____(4) Epidemiology
____(5) Genetics
____(6) Immunology
____(7) Microbiology
____(8) Molecular biology
____(9) Neurobiology
____(10) Plant biology
____(11) Pharmacology
____(12) Virology
____(13) Oncology
____(14) Other____________

NO POSTAGE
NECESSARY
IF MAILED
IN THE
UNITED STATES

BUSINESS REPLY MAIL
FIRST CLASS MAIL PERMIT NO. 150 HICKSVILLE, NY

POSTAGE WILL BE PAID BY ADDRESSEE

COLD SPRING HARBOR LABORATORY PRESS
10 SKYLINE DRIVE
PLAINVIEW, NY 11803-2500

NO POSTAGE
NECESSARY
IF MAILED
IN THE
UNITED STATES

BUSINESS REPLY MAIL
FIRST CLASS MAIL PERMIT NO. 150 HICKSVILLE, NY

POSTAGE WILL BE PAID BY ADDRESSEE

COLD SPRING HARBOR LABORATORY PRESS
10 SKYLINE DRIVE
PLAINVIEW, NY 11803-2500

NO POSTAGE
NECESSARY
IF MAILED
IN THE
UNITED STATES

BUSINESS REPLY MAIL
FIRST CLASS MAIL PERMIT NO. 150 HICKSVILLE, NY

POSTAGE WILL BE PAID BY ADDRESSEE

COLD SPRING HARBOR LABORATORY PRESS
10 SKYLINE DRIVE
PLAINVIEW, NY 11803-2500

Archaea

A LABORATORY MANUAL

F.T. Robb | A.R. Place
EDITOR-IN-CHIEF | MANAGING EDITOR
K.R. Sowers | H.J. Schreier
S. DasSarma | E.M. Fleischmann

Thermophiles

EDITED BY
F.T. Robb | A.R. Place

Cold Spring Harbor Laboratory Press 1995

Archaea

A LABORATORY MANUAL

Thermophiles

Printed in the United States of America
Design by Emily Harste

Back cover: Watercolor by Scott Fuqua. Representation of Shinkai 6500 from photograph courtesy of the Deep-Star group of JAMSTEC, Yokosuka, Japan.

The Library of Congress has cataloged the combined volume as follows:

Archaea : a laboratory manual / F.T. Robb, editor-in-chief ... [et al.].
p. cm.
Includes bibliographical references and index.
ISBN 0-87969-397-5
1. Archaebacteria--Research--Laboratory manuals. I. Robb, F.T. (Frank T.)
QR82.A69A73 1995
589.9--dc20 94-27420
CIP

Halophiles ISBN 0-87969-438-6
Methanogens ISBN 0-87969-439-4
Thermophiles ISBN 0-87969-440-8

All Cold Spring Harbor Laboratory Press publications may be ordered directly from Cold Spring Harbor Laboratory Press, 10 Skyline Drive, Plainview, New York 11803-2500. Phone: 1-800-843-4388 in Continental U.S. and Canada. All other locations: (516) 349-1930. FAX: (516) 349-1946.

CONTENTS

Contributors, vii
Contents of Companion Volumes, ix
Acknowledgments, xii
Preface, xiii
Foreword, xv

THERMOPHILIC ARCHAEA: An Overview, 3
M.W.W. Adams

GROWTH AND IDENTIFICATION

PROTOCOL 1
Isolation and Cultivation of Heterotrophic Hyperthermophiles from Deep-sea Hydrothermal Vents, 9
H.W. Jannasch, C.O. Wirsen, and T. Hoaki

PROTOCOL 2
Isolation, Growth, and Maintenance of Hyperthermophiles, 15
J.A. Baross

PROTOCOL 3
Plate Cultivation Technique for Strictly Anaerobic, Thermophilic, Sulfur-metabolizing Archaea, 25
G. Erauso, D. Prieur, A. Godfroy, and G. Raguénes

PROTOCOL 4
Continuous Culture Techniques for Extremely Thermophilic and Hyperthermophilic Archaea, 31
R.N. Schicho, S.H. Brown, I.I. Blumentals, T.L. Peeples, G.D. Duffaud, and R.M. Kelly

PROTOCOL 5
High-pressure–High-temperature Cultivation of Extremely Thermophilic Archaea, 37
C.M. Nelson and D.S. Clark

PROTOCOL 6
Large-scale Growth of Hyperthermophiles, 47
M.W.W. Adams

PROTOCOL 7
Culture of *Thermoplasma acidophilum*, 51
D. Searcy

BIOCHEMISTRY

PROTOCOL 8
Solubilization from Membranes and Purification of Hydrogenases from *Pyrodictium* spp., 63
R.J. Maier, T. Pihl, and J.B. Bingham

PROTOCOL 9
Purification of Hydrogenase and Ferredoxin from a Hyperthermophile, 67
M.W.W. Adams and Z.H. Zhou

PROTOCOL 10
Archaeal Lipid Analysis, 73
D.B. Hedrick and D.C. White

PROTOCOL 11
Glucose Metabolic Pathways in *Thermoplasma acidophilum* using ^{13}C- and ^{15}N-NMR Spectroscopy, 81
K.J. Stevenson, L.A. Manning, M.J. Danson, and D. McIntyre

MOLECULAR BIOLOGY AND GENETICS

PROTOCOL 12
Purification of Plasmids from Thermophilic and Hyperthermophilic Archaea, 87
F. Charbonnier and P. Forterre

PROTOCOL 13
Transfection of *Sulfolobus solfataricus*, 91
C. Schleper and W. Zillig

PROTOCOL 14
Preparation of Genomic DNA from Sulfur-dependent Hyperthermophilic Archaea, 95
V. Ramakrishnan and M.W.W. Adams

PROTOCOL 15
RNA Extraction from Sulfur-utilizing Thermophilic Archaea, 97
J. DiRuggiero and F.T. Robb

PROTOCOL 16
Reliable Amplification of Hyperthermophilic Archaeal 16S rRNA Genes by the Polymerase Chain Reaction, 101
A.-L. Reysenbach and N.R. Pace

PROTOCOL 17
Characterization of Archaeal Introns in the rRNA Genes of Hyperthermophiles, 107
J.Z. Dalgaard, J. Lykke-Andersen, and R.A. Garrett

PROTOCOL 18
Generation of Subtraction Probes for Isolation of Specific Genes in Thermophilic Archaea, 119
K.A. Robinson, F.T. Robb, and H.J. Schreier

PROTOCOL 19
Isolation of *Sulfolobus acidocaldarius* Mutants, 125
D. Grogan

PROTOCOL 20
Measurement of Mutation Rates in *Sulfolobus acidocaldarius*, 133
D. Grogan

PROTOCOL 21
Typing Marine Vent Thermophiles by DNA Polymerase Restriction Fragment Length Polymorphisms, 139
F.B. Perler, M.W. Southworth, D.G. Wilbur, and D. Wallace

PROTOCOL 22
Typing Hyperthermophilic Archaea Based on the 16S/23S rRNA Spacer Region, 149
J. DiRuggiero, J.H. Tuttle, and F.T. Robb

PROTOCOL 23
In Vitro Transcription from Natural and Mutated rDNA Promoters of the Extremely Thermophilic Archaeon *Sulfolobus shibatae*, 155
J. Hain and W. Zillig

APPENDICES

Appendix 1 Thermophilic Strains Available from DSM, 159
H. Hippe, B.J. Tindall, and M. Kracht

Appendix 2 Media for Thermophiles, 167
F.T. Robb and A.R. Place

Appendix 3 Chromosome Map of *Thermococcus celer*, 189
K.M. Noll

Appendix 4 Codon Usage Tables for Thermophilic Archaea, 191
J. DiRuggiero, K.M. Borges, and F.T. Robb

Appendix 5 The Berlin RNA Databank: Compilation of 5S rRNA and 5S rRNA Gene Sequences, 195
T. Specht and V.A. Erdmann

Appendix 6 16S and 23S rRNA-like Primers, 201
L. Achenbach and C.R. Woese

Appendix 7 Thermodynamic Data on the Activation and Denaturation of Proteins from Thermophilic Archaea, 205
H.H. Klump

Appendix 8 Suppliers, 209

Index, 211

CONTRIBUTORS

Frank T. Robb, Center of Marine Biotechnology, Baltimore, MD 21202
Allen R. Place, Center of Marine Biotechnology, Baltimore, MD 21202
Kevin R. Sowers, Center of Marine Biotechnology, Baltimore, MD 21202
Harold J. Schreier, Center of Marine Biotechnology, Baltimore, MD 21202
Shiladitya DasSarma, Department of Microbiology, University of Massachusetts, Amherst, MA 01003
Esther M. Fleischmann, Biological Sciences, University of Maryland, Baltimore County, Baltimore, MD 21228

Laurie Achenbach, Department of Microbiology, Southern Illinois University, Carbondale, IL 62901

Michael W.W. Adams, Department of Biochemistry, University of Georgia, Athens, GA 30602

John A. Baross, School of Oceanography, University of Washington, Seattle, WA 98195

James B. Bingham, Department of Biology, Johns Hopkins University, Baltimore, MD 21218

Ilse I. Blumentals, Life Technologies, Inc., Gaithersburg, MD 20884

Kimberly M. Borges, Department of Microbiology, University of Connecticut, Storrs, CT 06269

Stephen H. Brown, Novo-Nordisk Biotech, Inc., Davis, CA 95616

Franck Charbonnier, Institut de Génétique et Microbiologie, CNRS URA 1354, Université Paris Sud, F-91405 Orsay Cedex, France

Douglas S. Clark, Department of Chemical Engineering, University of California at Berkeley, Berkeley, CA 94720

Jacob Z. Dalgaard, Institute of Molecular Biology, University of Copenhagen, DK-1307 Copenhagen K, Denmark

Michael J. Danson, Department of Biochemistry, University of Bath, England

Jocelyne DiRuggiero, Center of Marine Biotechnology, University of Maryland System, Baltimore, MD 21202

Guy D. Duffaud, Department of Chemical Engineering, North Carolina State University, Raleigh, NC 27695-7905

Gaël Erauso, Mikrobiologisches Institut, ETH-Zentrum/LFV, CH-8092 Zürich, Switzerland

Volker Erdmann, Institut für Biochemie, Freie Universität Berlin, D-14195 Berlin, Germany

Patrick Forterre, Institut de Génétique et Microbiologie, CNRS URA 1354, Université Paris Sud, F-91405 Orsay Cedex, France

Roger A. Garrett, Institute of Molecular Biology, University of Copenhagen, DK-1307 Copenhagen K, Denmark

Anne Godfroy, Laboratoire de Biotechnologie, IFREMER Centre de Brest, 29680 Plouzané, France

Dennis W. Grogan, Department of Biological Sciences, University of Cincinnati, Cincinnati, OH 45221-0006

Johannes Hain, Bavarian Nordic Research Institute A/S, D-84764 Oberschleissheim, Germany

David B. Hedrick, Center for Environmental Biotechnology, University of Tennessee, Knoxville, TN 37923-2567

H. Hippe, DSM-Deutsche Sammlung von Mikroorganismen und Zellkulturen GmbH, D-38124, Braunschweig, Germany

Toshihiro Hoaki, Marine Biotechnology Institute, Shimizu, Shizuoka, 424-91 Japan

Holger W. Jannasch, Department of Biology, Woods Hole Oceanographic Institute, Woods Hole, MA 02543

Robert M. Kelly, Department of Chemical Engineering, North Carolina State University, Raleigh, NC 27695-7905

H.H. Klump, Department of Biochemistry, University of Cape Town, Rondebosch 7700, Republic of South Africa

Manfred Kracht, DSM-Deutsche Sammlung von Mikroorganismen und Zellkulturen GmbH, D-38124, Braunschweig, Germany

Jens Lykke-Andersen, Institute of Molecular Biology, University of Copenhagen, DK-1307 Copenhagen K, Denmark

Robert J. Maier, Department of Biology, Johns Hopkins University, Baltimore, MD 21218

Leslie A. Manning, Department of Biological Sciences, University of Calgary, Calgary T2N 1N4, Alberta, Canada

Deane McIntyre, Department of Biological Sciences, University of Calgary, Calgary T2N 1N4, Alberta, Canada

Chad M. Nelson, Advanced Fuel Research, Inc., East Hartford, CT 06108

Kenneth M. Noll, Department of Microbiology, University of Connecticut, Storrs, CT 06269

Norman R. Pace, Biology Department, Indiana University, Bloomington, IN 47405

Tonya L. Peeples, Department of Chemical and Biochemical Engineering, University of Iowa, Iowa City, IA 52242

Francine B. Perler, New England Biolabs, Inc., Beverly, MA 01915

Todd Pihl, Department of Biology, Johns Hopkins University, Baltimore, MD 21218

Allen R. Place, Center of Marine Biotechnology, Baltimore, MD 21202

Daniel Prieur, Mikrobiologisches Institut, ETH-Zentrum/LFV, CH-8092 Zürich, Switzerland

Gérard Raguenes, Laboratoire de Biotechnologie, IFREMER Centre de Brest, BP 70, 29680 Plouzané, France

Vijayaragavan Ramakrishnan, Department of Biochemistry, University of Georgia, Athens, GA 30602

Anna-Louise Reysenbach, Biology Department, Indiana University, Bloomington, IN 47405

Frank T. Robb, Center of Marine Biotechnology, Baltimore, MD 21202

Kelly A. Robinson, University of Maryland, Baltimore County, MD 21229 and Center of Marine Biotechnology, Baltimore, MD 21202

Richard N. Schicho, Warner-Lambert Co., Morris Plains, NJ 07950

Christa Schleper, The Agouron Institute, La Jolla, CA 92037

Harold J. Schreier, Department of Biology, University of Maryland, Baltimore County, MD 21229 and Center of Marine Biotechnology, Baltimore, MD 21202

Dennis G. Searcy, Biology Department, University of Massachusetts, Amherst MA 01003-5810

Maurice W. Southworth, New England Biolabs, Inc., Beverly, MA 01915

Thomas Specht, Institut für Biochemie, Freie Universität Berlin, D-14195 Berlin, Germany

Kenneth J. Stevenson, Department of Biological Sciences, University of Calgary, Calgary T2N 1N4, Alberta, Canada

Brian J. Tindall, DSM-Deutsche Sammlung von Mikroorganismen und Zellkulturen GmbH, D-38124, Braunschweig, Germany

Jon H. Tuttle, Chesapeake Biological Laboratory, Center for Environmental and Estuarine Studies, University of Maryland System, Solomons, MD 20688

David Wallace, New England Biolabs, Inc., Beverly, MA 01915

David C. White, Center for Environmental Biotechnology, University of Tennessee, Knoxville, TN 37923-2567

Dean G. Wilbur, New England Biolabs, Beverly, MA 01915

Carl O. Wirsen, Department of Biology, Woods Hole Oceanographic Institute, Woods Hole, MA 02543

Carl R. Woese, Department of Microbiology, University of Illinois, Urbana, IL 61801

Zhi Hao Zhou, Department of Biochemistry, University of Georgia, Athens, GA 30602

Wolfram Zillig, Max-Planck-Institut für Biochemie, D-8033 Martinsried, Germany

Contents of Companion Volumes

Halophiles

EDITED BY
S. DasSarma and E.M. Fleischmann

Halophilic Archaea: An Overview
Cultivation of Halophilic Archaea
Growth of Halophilic Archaea Under Conditions of Low Oxygen Tension and High Light Intensity
Selection and Screening Methods for Halophilic Archaeal Rhodopsin Mutants
Electron Microscopy of Halophilic Archaea
Isoprenoids and Polar Lipids of Extreme Halophiles
Isolation of Purple Membranes
Purification of Halorhodopsin from *Halobacterium salinarium*
Spectroscopic Assays for Sensory Rhodopsins I and II in *Halobacterium salinarium* Cells and Enriched Membrane Preparations
Purification of the ATPase from *Halobacterium saccharovorum*
Electrophoresis of *Halobacterium* Gas Vesicle Protein in Phenol-Acetic Acid-Urea (PAU) Gels
Isolation of Halophilic Archaeal Ferredoxin
Halophilic Proteases from Halophilic Archaea
Purification of Halophilic Archaeal Ketohexokinase from *Haloarcula vallismortis*
Purification of Superoxide Dismutase and a Catalase-Peroxidase in a High-salt Environment
Purification of *Halobacterium halobium* DNA Polymerases
Purification of DNA-dependent RNA Polymerase from *Halobacterium halobium*
Preparation of Transfer RNA, Aminoacyl-tRNA Synthetases, and tRNAs Specific for an Amino Acid from Extreme Halophiles
Cell-free Protein Synthesis System for Halophilic Archaea
Isolation of Free and Membrane-bound Polysomes from *Halobacterium halobium*
Total Reconstitution of Ribosomal Subunits of *Haloferax mediterranei*
The MPD-NaCl-H_2O System for the Crystallization of Halophilic Proteins
Purification and Labeling of RNA from *Halobacterium halobium* for Identification of Regulated Genes
A Rapid Procedure for the Isolation of RNA from *Haloferax volcanii*
Analysis of RNA Transcripts in Archaea

Isolation of Genomic and Plasmid DNAs from *Halobacterium halobium*
Preparation of Intact, Agarose-embedded DNA from *Halobacterium halobium* and Its Digestion by Restriction Enzyme
Fractionation of Halophilic Archaeal DNA into FI and FII Using Affinity Chromatography on Malachite Green Bisacrylamide
Electrophoretic Mobility Shift Assays with Crude Extracts of *Halobacterium halobium*
Transformation of Halophilic Archaea
Gene Replacement in *Halobacterium halobium*
Selection for Spontaneous and Engineered Mutations in the rRNA Genes in Halophilic Archaea
Appendices

Methanogens

EDITED BY
K.R. Sowers and H.J. Schreier

Methanogenic Archaea: An Overview
Techniques for Anaerobic Growth
Growth of Methanogens on Solidified Medium
Plating Techniques for Extremely Thermophilic Methanogens
Growth of *Methanosarcina* spp. as Single Cells
Large-scale Culture Techniques for Methanogenic Archaea
Large-scale Growth of Methanogens That Utilize Acetate and Formate in a pH Auxostat
Short- and Long-term Maintenance of Methanogen Stock Cultures
Long-term Maintenance of Methanogen Stock Cultures in Glycerol
Techniques for Monitoring Methanogen Cell Growth
Immunologic Identification and Analysis of Methanogenic Archaea
Use of Fluorescent Probes for Determinative Microscopy of Methanogenic Archaea
Electron Microscopy Techniques for the Archaea
Techniques for Anaerobic Biochemistry
Purification of H_4MPT-dependent Enzymes of Methanogenesis
Purification of the Methylreductase System and H_2:Heterodisulfide Oxidoreductase Complex from *Methanobacterium thermoautotrophicum*
Purification of Coenzymes from the Methanogenic Pathway
Purification of the Coenzyme F_{420}-reducing Hydrogenase from *Methanobacterium formicicum*
Purification of Formate Dehydrogenase from *Methanobacterium formicicum*
Purification of *Methanosarcina thermophila* Acetate Kinase and Phosphotransacetylase Overproduced in *Escherichia coli*
Purification of Carbon Monoxide Dehydrogenase from *Methanosarcina thermophila*
Purification of the Carbon Monoxide Dehydrogenase-linked Ferredoxin from *Methanosarcina thermophila*
Purification of Carbonic Anhydrase from *Methanosarcina thermophila*
Purification of Methanol-catabolizing Enzymes from Methanogenic Archaea

Purification of Enzymes Involved in Alcohol Utilization by Methanogenic Archaea
Electron Microscopy Technique for Immunocytochemical Localization of Enzymes in Methanogenic Archaea
Purification of Nitrogenase from the Methanogenic Archaeon *Methanosarcina barkeri*
Preparation of Pseudomurein Endopeptidase from *Methanobacterium wolfei*
Purification of a Disaggregating Enzyme for Generating *Methanosarcina* Single Cells
Purification of DNA-dependent RNA Polymerases from Methanogens
Purfication of the Histone HMf from *Methanothermus fervidus*
Purification of the Flagellins from Methanogenic Archaea
Isolation and Analysis of Cell Walls from Methanogenic Archaea
Purification of Ether Lipids and Liposome Formation from Polar Lipid Extracts of Methanogenic Archaea
Turnover of Amino Acids in Methanogens Monitored by NMR Spectroscopy
Detection of Phosphorylated Small Molecules in *Methanobacterium thermoautotrophicum* Strain ΔH by ^{31}P- and ^{13}C-NMR Spectroscopy
Extraction and Detection of Compatible Intracellular Solutes
Isolation of Chromosomal and Plasmid DNAs from Methanogenic Archaea
Isolation of RNA from Methanogenic Archaea
Chromosomal Mapping of Methanogenic Archaea
Gene Cloning following Size Selection of Restriction Fragments
Gene Selection and Cloning with Bacteriophage λ Libraries
Mutagenesis of *Methanococcus* spp. with Ethylmethanesulfonate
Selection for Auxotrophic Mutants of *Methanococcus*
Integration Vectors for Methanococci
Transformation of *Methanococcus voltae* by Protoplast Regeneration
Transduction of *Methanobacterium thermoautotrophicum* Marburg
Techniques for In Vitro Transcription Assays with *Methanococcus thermolithotrophicus*
Appendices

ACKNOWLEDGMENTS

We express our thanks for the generous support provided by the following sponsors of this project: The U.S. Office of Naval Research, United States Biochemical Corporation, Entotech, Genencor, and New England Biolabs. We are also grateful for additional funding provided by Promega Corporation. We would like to express our appreciation to the University of Maryland Biotechnology Institute, the National Science Foundation, and the Department of Energy for their support as well.

PREFACE

This manual came into being as the result of a workshop held at the Center for Marine Biotechnology in July, 1990. The workshop reflected the unique phylogeny and extraordinary metabolic properties of a group of organisms that was then called the Archaebacteria. It also reflected the awkward nature of the beasts themselves. The Archaea, as they are now called, contain the most oxygen-sensitive anaerobes, the most extreme thermophiles, and halophiles whose salt-dependent enzymes require a rethinking on the methods of protein biochemistry. It is clear that the factor limiting archaeal research was the need for a central, comprehensive source of specialized methods. A group at the Center of Marine Biotechnology—Al Place, Hal Schreier, Esther Fleischmann, and Frank Robb—responded to this need by producing a small technical manual consisting of contributions from the participants at the workshop. Our original manual, funded by the U.S. Office of Naval Research and printed at the Center of Marine Biotechnology, provided a sampler of experimental methods.

Jeffrey Miller prompted the interest of Cold Spring Harbor Laboratory Press in the project, which led to the assembly of the present set of protocols, with overviews of the halophiles, methanogens, and thermophiles and a number of appendices. Carl Woese, whose studies on the small ribosomal subunit RNA first established that the Archaea are organisms very different from bacteria, emphatically describes the battle to establish this group in their rightful position as "honorary eukaryotes" with prokaryotic cell organization. This goes against the grain, so to speak, of the dogma that prokaryotes are monophyletic. Because of this, the Archaea have suffered through successive awkward names, such as Archaebacteria, Archeobacteria, Archeote, and the like.

There are signs that the apprenticeship of the group as a microbial sideshow is about to end. The Archaea recently received a significant salute through the choice of two methanogens (*Methanothermus fervidus* and *Methanobacterium thermoautotrophicum*) and a sulfur-reducing hyperthermophile, *Pyrococcus furiosus*, as organisms targeted for genomic sequencing by the Microbial Genomes Initiative funded by the U.S. Department of Energy. The sulfur-oxidizing thermophile, *Sulfolobus solfataricus*, has been chosen by the Canadian Center for Advanced Studies for a 3-year genomic sequencing program. Another new development is the unexpected emergence of Archaea as a major component of the prokaryotic biomass in oceanic waters. Three "sightings" of previously unrecognized archaeal populations in Antarctic and Alaskan waters, where the Archaea can constitute up to 30% of the submicrometer population, are summarized by Gary Olsen in a *Nature* (*371:* 657–658 [1994] News and Views) article. These enigmatic cells, which may be major players in global microbial ecology, are unknown in terms of their physiology, since none have been cultured. This underlines the need for specialized methods for isolation and culturing. In the midst of this battle for recognition, it is clear that the Archaea isolated thus far harbor unique molecular adaptations that are tailor-made for applications in molecular biology and biotechnology, such as thermostable DNA polymerases for PCR, methanogenesis as an alternative fuel source, and bacteriorhodopsin for production of computer chips. Knowing the high potential that Archaea have for both basic and applied science, it is our hope that this manual will provide a practical user guide for new and established investigators in archaeal research.

As Editors, we owe a great deal to the initial contributors to the workshop and the first manual. All of the authors have provided outstanding copy and have shown patient attention to detail. The archaeal research community possesses an unrivaled level of cooperation, and without it, this manual could not have been completed. The strong response provided CSHL Press with a far larger publishing job than they had anticipated. The artistic talents of Scott Fuqua evident throughout the manuals are greatly appreciated. We have also had generous sponsorship from five organizations, as detailed in the Acknowledgments, particularly from the U.S. Office of Naval Research which has supported the project and the Center of Marine Biotechnology from the outset. The University of Maryland is also to be thanked for housing all of the editors, except Shil DasSarma, who is at the University of Massachusetts at Amherst.

Finally, we wish to thank John Inglis, Nancy Ford, Dotty Brown, Mary Cozza, and Susan Schaefer of Cold Spring Harbor Laboratory Press. Their skill and professional acumen is truly impressive.

Frank Robb
Al Place
Kevin Sowers
Hal Schreier
Shil DasSarma
Esther Fleischmann

February 5, 1995

FOREWORD: When Is a Prokaryote Not a Prokaryote?

The Archaea came out of nowhere—at least out of a conceptual vacuum—suddenly in late 1977 (Fox et al. 1977; Woese and Fox 1977). Biology had lived innocently for decades, in a sort of intellectual Garden of Eden, where it was implicitly (and dogmatically) assumed that at the highest genealogical level, the living world comprised only two types of self-replicating entities: prokaryotes and eukaryotes (Murray 1974). Yet there they were, the Archaea, an unexpected "third form of life," and biologists had no place for them in their world.

It is important to understand why biologists reacted with incredulity to the discovery of the Archaea—and not merely for reasons of historical interest. The conceptual climate of 1977 remains in essence the conceptual climate of today. The eukaryote/prokaryote dichotomy still dominates biology and still influences in particular our perception of the Archaea. Microbiologists tend to remain fixed in their belief that all prokaryotes are "basically alike." Eukaryotic biologists still find it difficult to accept that studying some group of prokaryotes might shed light on the specific nature of the eukaryotic cell. Only when we stop looking at the living world in this unproductive bimodal way will we come to see the Archaea in their true light and study them correctly.

The source of biology's problem here was that it did not understand the true nature of the prokaryote/eukaryote dichotomy. The concept had been defined in two stages: an initial cytological definition and a subsequent molecular one. By its initial definition, the prokaryote was merely a cell that did not possess this or that eukaryotic (cellular) feature, in particular a nuclear membrane. The group was therefore in essence negatively defined and so was phylogenetically meaningless. The subsequent molecular redefinition, in which homologous macromolecules could be used to define and distinguish prokaryote and eukaryote, did convey phylogenetic meaning. But that redefinition was never properly done. Biologists of the 1950s and 1960s naively accepted the initial cytological definition of the prokaryote as phylogenetically valid, and so the molecular redefinition of "prokaryote" was fashioned for all intents and purposes using one "representative" species, *Escherichia coli.* It is this failure to understand that too much had been taken for granted, that the assumed monophyletic nature of the prokaryote had never been tested, which laid the ground work for biology's incredulity at the discovery of the Archaea.

The discovery of the Archaea has opened our eyes to the pernicious nature of the prokaryote/eukaryote dichotomy. The prokaryote/eukaryote notion deemphasized the essence of microbiology— microbial diversity. It has also served to divide biology falsely into two disparate camps, separated methodologically, conceptually, academically, in terms of scientific funding and in general scientific outlook. The prokaryote/eukaryote view has also deemphasized evolutionary matters, the Darwinian principle that an organism's evolutionary history is part and parcel of the organism; and it has made us complacent about the deep and interesting questions concerning the origin and nature of the eukaryotic cell, and the ancestor of all extant life.

This is the climate—the world of the prokaryote/eukaryote and the deemphasis of evolution inherent in the molecular world view—into which the Archaea were born, as an unanticipated and unwanted offspring. It has been altogether too long an uphill struggle to get microbiologists (and biologists) to see the false assumption upon which the prokaryote/eukaryote dichotomy rests, and so to release their prejudice that because they lack nuclear membranes, the Archaea are necessarily related to the bacteria. But, as this manual bears witness, the tide is beginning to turn, if not because biologists have thought through the flaw in the prokaryote/eukaryote dichotomy, then because the Archaea are compellingly interesting in their own right. Let us not forget, however, that the number of biologists now actively interested in the Archaea remains small, especially from the perspective of the group's primary taxonomic rank: The old prokaryote/eukaryote world view still holds sway, still tends to blind microbiology to the significance of the Archaea.

Microbiology today is at a critical juncture. It is tending, on the one hand, to become an increasingly practical discipline, focused on medical and agricultural problems and the like; microorganisms are increasingly seen as systems mainly useful for genetic engineering. Yet the sequence revolution and the existence for the first time of a phylogenetic framework within which to operate, to relate and understand facts, tends to pull microbiology in another direction, toward evolutionary considerations, renewed interest in microbial diversity, and the study of microbial ecology for its own sake. It is the Archaea that are helping to draw us in this latter direction, awakening us to fundamental biological problems and reinvigorating microbiology. The universal phylogenetic tree has now become a reality. The problem of the universal ancestor, cast in a new light, is gaining interest and now appears (to some extent) tractable. The role of prokaryotes in the origin of the eukaryotic cell has taken on new dimensions and new promise. The origin and early evolution of life are moving out of the arena of speculation and chemical experimentation of questionable relevance. Testable theories are emerging that some day may dovetail with the study of the evolution of microbial (biochemical) diversity. Microbial ecology has finally come into its own, with a new evolutionary perspective that draws strength from the introduction of the molecular phylogenetic techniques that led to the discovery of the Archaea.

We should study the Archaea for what they are, for what they naturally show us. The archaeal domain comprises (at least) two major kingdoms, the Euryarcheota and the Crenarchaeota. The nature of this bipartite split needs to be understood. Comparative studies of all kinds are definitely indicated. The Archaea (particularly the Crenarchaeota) are basically thermophilic. They are systems par excellence for understanding (and utilizing) high-temperature biology. Methanogens hold the key to the production of methane on this planet. And the extreme halophiles will necessarily provide insights into life and catalysis under saturating salt conditions. On the molecular level, the Archaea increasingly begin to resemble the Eukarya: in transcription signals, transcription factors, chaperones, and histones. Archaea are not bacteria. Their kinship is with the eukaryotes (albeit distant), and they should be used to understand the nature and evolution of the eukaryotic cells. However, above all, they are Archaea, a unique domain of organisms to be studied simply for that reason. Good hunting!

C.R. Woese

REFERENCES

Fox, G.E., L.J. Magrum, W.E. Balch, R.S. Wolfe, and C.R. Woese. 1977. Classification of methanogenic bacterial by 16S ribosomal RNA characterization. *Proc. Natl. Acad. Sci.* **74:** 4537–4541.

Murray, R.G.E. 1974. A place for bacteria in the living world. In *Bergey's manual of determinative biology*, 8th edition (ed. R.E. Buchanan and N.E. Gibbons), pp. 4–9. Williams and Wilkins, Baltimore.

Woese, C.R. and G.E. Fox. 1977. Phylogenetic structure of the prokaryotic domain: The primary kingdoms. *Proc. Natl. Acad. Sci.* **74:** 5088–5090.

Thermophiles

Artist's impression of the submersible Shinkai 6500 exploring hydrothermal vents in the Okinawa Trough.

THERMOPHILIC ARCHAEA: An Overview

M.W.W. Adams

Thermophilic Archaea encompass three of the most seminal developments that have occurred so far in the field of microbiology. All three took place in the late 1970s and early 1980s: (1) the establishment of Archaea (Archaebacteria) as a new domain of life (Woese 1987; Woese et al. 1990); (2) the discovery of deep-sea hydrothermal vents with dense populations of microorganisms flourishing nearby (Jannasch and Mottl 1985); and (3) the isolation from shallow marine volcanic vents of so-called "hyperthermophilic" microorganisms growing at temperatures near and above 100°C (Stetter 1986). Now, a decade or so later, a variety of different genera are known that grow above 90°C; many of them have been isolated from deep-sea hydrothermal environments, and almost all are classified as Archaea (Stetter et al. 1990; Adams 1993).

In fact, when the Archaea were first established as a distinct life-form in the early 1980s, they included methanogens, extreme halophiles, and some thermophilic organisms growing at approximately 70°C (species of *Sulfolobus* and *Thermoplasma*). This division of the Archaea into three distinct types of organism still holds true today, as those that are not classified as methanogenic or extremely halophilic are all thermophilic to a greater or lesser extent. This might change, however, with the recent identification of novel marine Archaea growing at 5–15°C (Olsen 1994). As shown in Figure 1, virtually all of the thermophilic Archaea have been isolated in just the last few years. Unfortunately, thermophily, like beauty, is somewhat in the eye of the beholder, and terms such as thermophilic, extremely thermophilic, and hyperthermophilic have yet to be clearly defined. Nevertheless, as illustrated in Figure 1, it is clear that a quantum leap in "thermophily" occurred in the early 1980s, and since then about 20 genera have been isolated that are able to grow optimally at or above 80°C. Also indicated in Figure 1 is the fact that all but two of these genera (*Thermotoga* and *Aquifex*; see Stetter et al. 1990; Adams 1993) are classified within the Archaea. In contrast, a variety of organisms are known that grow optimally in the range 50–70°C, and almost all of these are bacteria. It therefore seems appropriate to make some distinction between these two groups.

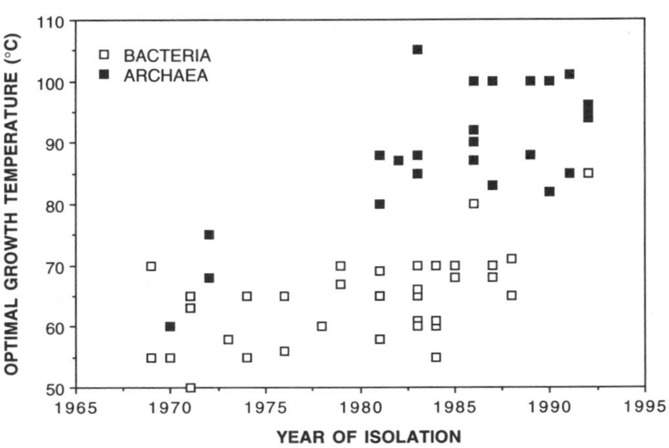

Figure 1 Isolation of Thermophilic Microorganisms. (Modified from Adams 1992.)

Similarly, in considering organisms that grow near 100°C, the ability to grow at or above 90°C has been used to classify an organism as a "hyperthermophile" (Adams 1993). This definition includes most, although not all, of the organisms with optimum growth temperatures above 80°C (Fig. 1). Hence, with only a few exceptions, the thermophilic Archaea can be considered as hyperthermophiles.

The currently known thermophilic Archaea and some of their properties are listed in Table 1. They can be divided into three groups. The majority fall into the "sulfur-dependent" category, as they obtain energy for growth primarily by the metabolism of elemental sulfur (S^o). The second group contains a unique sulfate-reducing genus, *Archaeoglobus*, which is also hyperthermophilic. The third category consists of thermophilic methanogens, of which there are three genera that can be classified as hyperthermophiles (Table 1). It should be noted that several other methanogens are known that grow at temperatures up to 60°C or so. Thus, the majority of the thermophilic Archaea are S^o-dependent organisms, and these are the main focus of the following series of protocols.

The S^o-dependent Archaea can be further subdivided into two main groups: anaerobic S^o-reducers, which grow at and above 90°C, and aerobic S^o-oxidizers, which are generally less thermophilic. The first category includes the Thermoproteales, Thermococcales, and several as yet unclassified organisms. These are strictly anaerobic heterotrophs that are obligately dependent on the reduction of S^o to H_2S for optimal growth (Table 1). In fact, of these organisms, only species of *Pyrococcus, Thermococcus,* and *Hyperthermus* show significant growth in the absence of S^o. Most of these anaerobic heterotrophs utilize only complex peptide mixtures such as yeast and meat extracts as carbon and nitrogen sources. Only a few of these organisms metabolize carbohydrates, including starch, glycogen, and maltose, but they also require peptides as a nitrogen source. As indicated in Table 1, some of these S^o-dependent heterotrophs are able to grow autotrophically, using H_2 as the electron donor for S^o reduction. All of these organisms are able to grow at 90°C and above. Most are of marine origin and several have been isolated near deep-sea vents. Only members of the Thermoproteaceae have been found in continental hot springs.

The second category of S^o-dependent Archaea include species of the Sulfolobales (Table 1). In contrast to the S^o-reducing heterotrophs, these are typically acidophilic aerobes that obtain energy for growth by the oxidation of S^o to sulfuric acid. Remarkably, species of *Acidianus* and *Desulfurolobus* also grow under anaerobic conditions by the reduction of S^o to H_2S using H_2 as the electron donor. *Stygiolobus* is unique among the Sulfolobales as it does not grow under aerobic conditions. As shown in Table 1, some species within the Sulfolobales are facultative autotrophs and are able to grow in complex media containing yeast or meat extracts. The Sulfolobales are also less thermophilic than the Thermoproteales and Thermococcales, with only species of *Acidianus* being able to grow at or above 90°C. In addition, they mainly inhabit continental sulfur-rich springs, although some species are also found near shallow marine volcanic vents. This category of S^o-dependent Archaea also includes the unique genus *Thermoplasma,* which occupies an isolated position in archaeal phylogeny (Woese 1987). Species of *Thermoplasma* are facultatively anaerobic heterotrophs. They grow optimally near 60°C both with and without S^o and can utilize monosaccharides as a carbon source.

In considering the effects of temperature on growth physiology, it is apparent that at the present upper temperature limits of life, the predominant metabolism is strictly anaerobic, heterotrophic S^o reduction. In addition, almost all of the hyperthermophilic species require complex organic mixtures as carbon and nitrogen sources, although a few species (including the methanogens) are able to grow autotrophically with H_2 as the electron donor. The ability to use O_2 as a terminal electron acceptor is very limited at temperatures above 90°C. In contrast, at slightly lower temperatures, it is the aerobic, S^o-oxidizing autotrophs that predominate. On the other hand, methanogenesis is a seemingly unique metabolic mode, spanning the mesophilic to hyperthermophilic temperature range. Habitat also plays a role in defining

Table 1 Thermophilic and Hyperthermophilic Archaea

Order (Family) Genus	T_{max}[a]	Physiology[b]	Donors[c]	Acceptors[c]	Habitat[d]
SULFUR-DEPENDENT ARCHAEA[e]					
Thermoproteales (Thermoproteaceae)					
Pyrobaculum	102°	hetero (auto)	org N (H_2)	S°	c
Thermofilum	100°	hetero	org N	S°	c
Thermoproteus	92°	hetero (auto)	org C,N (H_2)	S°	c
Thermoproteales (Desulfurococcaceae)					
Pyrodictium	110°	hetero (auto)	org C,N (H_2)	S°	m/d
Staphylothermus	98°	hetero	org N	S°	m/d
Thermodiscus	98°	hetero	org N	S°	m
Desulfurococcus	90°	hetero	org N	S°	m/d
Thermococcales					
Pyrococcus	105°	hetero	org C,N	S° (–)[f]	m
Thermococcus	97°	hetero	org C,N	S° (–)	m/d
(Unclassified)					
Hyperthermus	110°	hetero	org N (H_2)	S° (–)	m
"ES-4"	108°	hetero	org C,N	S°	d
"GB-D"	103°	hetero	org N	S°	d
"GE-5"	102°	hetero	org N	S°	d
"ES-1"	91°	hetero	org C,N	S°	d
Sulfolobales					
Acidianus	96°	auto	S°, H_2	O_2 (S°)	m/c
Sulfolobus	87°	auto	S°, H_2 (org C,N)	O_2 (S°)	c
Desulfurolobus	87°	auto	S°, H_2	O_2 (S°)	m/c
Stygiolobus	88°	auto	H_2	S°	c
Metallosphaera	80°	auto	S° (org N)	O_2	c
Thermoplasmatales					
Thermoplasma	67°	hetero	org C,N	(–), O_2, S°	c
SULFATE-REDUCING ARCHAEA					
Archaeoglobus	95°	hetero (auto)	org C (H_2)	SO_4, S_2O_3	m/d
METHANOGENIC ARCHAEA					
Methanococcus	91°	auto	H_2	CO_2	m/d
Methanothermus	97°	auto	H_2	CO_2	c
Methanopyrus	110°	auto	H_2	CO_2	m/d

[a] Maximum growth temperature.
[b] Indicates whether species of a genus are heterotrophs (hetero) or autotrophs (auto) or both.
[c] Electron donors and acceptors.
[d] Isolated from continental (c), shallow marine (m), or deep-sea (d) geothermal areas.
[e] The sulfur-dependent genera are grouped in separate orders, except for *Hyperthermus,* ES-1, ES-4, GB-D, and GE-5, which have yet to be classified.
[f] (–) indicates growth in the absence of S°. (Modified from Adams 1993.)

physiology and thermophily (Table 1). Continental geothermally heated springs and geysers do not exceed 100°C and are typically acidic due to the aerobic oxidation of sulfide, although subsurface water not exposed to oxygen can be neutral to alkaline. By comparison, marine geothermal vents are extensively mineralized because of the high reactivity of saline water and are typically highly reducing, anaerobic, and only slightly acidic. Moreover, because of the hydrostatic pressure in deep-sea locales, the water does not boil and temperatures can range up to 400°C. Hence, most of the strictly anaerobic, heterotrophic S^0-reducing hyperthermophiles are of marine origin, whereas the less thermophilic, aerobic S^0-oxidizers are generally found in continental ecosystems.

The practical problems of isolating, identifying, and growing species of thermophilic Archaea are obvious, and the first section, Growth and Identification, is devoted to these aspects. Isolation of hyperthermophiles is particularly challenging because, in addition to gaining access to their natural environments, the use and maintenance of high temperatures, high hydrostatic and/or gas pressures, anaerobic conditions, and highly acidic or highly reducing media are all factors that may have to be considered. Such procedures are described in detail in Protocol 1 (H.W. Jannasch et al.) and Protocol 2 (J.A. Baross). Once isolation and identification of hyperthermophilic organisms are accomplished, however, many obstacles must be overcome to successfully culture them reproducibly on either a small or large scale. Use of conventional plating media is precluded simply by the need for high temperatures, and this is usually complicated further by a requirement for anaerobicity. G. Erauso et al. (Protocol 3) describe how such problems can be overcome, even for hyperthermophilic anaerobes.

Attempts to grow many hyperthermophiles have also been severely hindered by their need to reduce S^0 for optimal growth and by the fact that they are of marine origin. Hence, because of the production of corrosive and toxic H_2S in high-salt-containing media, the routine cultivation of obligate S^0-reducing organisms in conventional stainless steel vessels is prohibited. Accurate temperature maintenance can also be problematic. Protocol 4 (R.N. Schicho et al.) describes continuous and batch culture techniques using glass vessels for heterotrophic and autotrophic hyperthermophiles. The effect of pressure is an often overlooked parameter in growing hyperthermophiles, in particular those from deep-sea environments. Although an obligately barophilic archaeon has yet to be isolated, pressure does modify the growth rates and cell yields of some hyperthermophiles. Protocol 5 (C.M. Nelson and D.S. Clark) describes the specialized equipment needed to investigate this. The procedures used in the author's laboratory for the large-scale (600-liter) growth of hyperthermophiles are given in Protocol 6 (M.W.W. Adams). Protocol 7 (D. Searcy) describes the unusual conditions required to cultivate one of the most unique members of the thermophilic Archaea, *Thermoplasma*, as developed by one of the pioneers in its biochemical characterization.

Because the thermophilic Archaea are such a recent discovery, and in most cases cannot be grown by conventional plating and fermentation techniques, our understanding of their biochemistry, genetics, and molecular biology is still at an early stage. Nevertheless, a number of enzymes have been purified from the hyperthermophiles (Adams 1993), and the heterotrophs have been shown to contain unusual pathways for carbohydrate and peptide metabolism (Kelly and Adams 1994). In the Biochemistry Section, the purification of hydrogenase is described from an H_2-oxidizing species of *Pyrodictium* in Protocol 8 (R.J. Maier et al.). Protocol 9 (M.W.W. Adams and Z.H. Zhou) describes the purification of hydrogenase and ferredoxin from an H_2-evolving species of *Pyrococcus*. The latter also includes techniques needed for the large-scale purification of oxygen-sensitive enzymes. The unique membrane lipids of the Archaea still remain one of their most distinctive features from an analytical perspective, and Protocol 10 (D.B. Hedrick and D.C. White) discusses the methods involved. Protocol 11 (K.J. Stevenson et al.) describes the use of NMR spectroscopy to investigate metabolic pathways in a species of *Thermoplasma*.

Aspects of the genetics of the thermophilic Archaea occupy the remaining protocols. In most cases, the procedures that are described have only recently been devised, which reflects the difficulties in overcoming the special problems of dealing with these organisms. It will be apparent to the reader that the molecular biology of the hyperthermophilic organisms in particular is much less developed than the techniques to isolate them and to study their biochemistry. Hence, procedures to isolate genomic DNA (Protocol 14, V. Ramakrishnan and M.W.W. Adams), total RNA (Protocol 15, J. DiRuggiero and F.T. Robb), plasmids (Protocol 12, F. Charbonnier et al.), and specific genes (Protocol 18, K.A. Robinson et al.) from hyperthermophiles are described. The characterization of these organisms as known or novel isolates is also problematic, and techniques are presented to assist in this at the nucleotide level. These include analyses using 16S rRNA sequences (Protocol 16, A.-L. Reysenbach and N.R. Pace), 16S/23S rRNA RFLPs (Protocol 22, J. DiRuggiero et al.), and DNA polymerase RFLPs (Protocol 21, F.B. Perler et al.). Another unique feature of archaeal rRNA is the presence of introns, and their characterization is discussed in Protocol 17 (J.Z. Dalgaard et al.).

The recent breakthroughs in the molecular genetics of species of *Sulfolobus* are the topics of the remaining four protocols, the only thermophilic Archaea for which mutants and genetic transfer techniques are available. The subjects are transfection (Protocol 13, C. Schleper and W. Zillig), mutant isolation and measurements of mutation rates (Protocols 19 and 20, D.W. Grogan), and in vitro transcription using promoters of rRNA genes (Protocol 23, J. Hain and W. Zillig).

The Appendices contain a wealth of additional information on the thermophilic Archaea. Included are the descriptions of strains available from DSM, media recipes for cell culturing, a physical map of *Thermococcus celer*, and information on codon usage, 5S RNA and gene sequences, 16S and 23S rRNA-like primer sequences, and thermodynamic values for some proteins from thermophilic Archaea.

Our knowledge of the thermophilic Archaea will no doubt expand considerably in the next few years as each of the many fascinating aspects of this unique group of organisms is developed further. They have already revolutionized the fundamental way in which life-forms are classified and have extended the limits of the conditions under which life can thrive. They are providing a new insight into evolution and clues to the mechanisms by which life may have first originated and higher life-forms may have developed. They harbor unusual metabolic pathways containing some new enzymes and have new versions of known enzymes but with unprecedented thermal stability. The protocols described herein will be of great use to researchers both within the field and to those who might considering entering it. One could even hope that the use of these techniques might enable the isolation of organisms that are even more thermophilic or ones that are even more slowly evolving. Indeed, perhaps such life-forms, if they (still) exist, must be one and the same.

REFERENCES

Adams, M.W.W. 1992. Novel iron sulfur clusters in metalloenzymes and redox proteins from extremely thermophilic bacteria. *Adv. Inorg. Chem.* **38:** 341–396.

———. 1993. Enzymes and proteins from organisms that grow near and above 100°C. *Annu. Rev. Microbiol.* **47:** 627–658.

———. 1994. Biochemical diversity among sulfur-dependent hyperthermophilic microorganisms. *FEMS Microbiol. Rev.* (in press).

Olsen, G.T. 1994. Archaea, archaea everywhere. *Nature* **371:** 657–658.

Jannasch, H.W. and M.J. Mottl. 1985. Geomicrobiology of deep-sea hydrothermal vents. *Science* **229:** 717–725.

Stetter, K.O. 1986. Diversity of extremely thermophilic archaebacteria. In *The thermophiles, general, molecular and applied microbiology* (ed. T.D. Brock), pp. 39–74. John Wiley, New York.

Stetter, K.O., G. Fiala, G. Huber, R. Huber, and G. Segerer. 1990. Hyperthermophilic microorganisms. *FEMS Microbiol. Rev.* **75:** 117–124.

Woese, C.R. 1987. Bacterial evolution. *Microbiol. Rev.* **51:** 221–271.

Woese, C.R., O. Kandler, and M.L. Wheelis. 1990. Towards a natural system of organisms: Proposal for the domains of Archaea, Bacteria and Eucarya. *Proc. Natl. Acad. Sci.* **87:** 4576–4579.

PROTOCOL 1

Isolation and Cultivation of Heterotrophic Hyperthermophiles from Deep-sea Hydrothermal Vents

H.W. Jannasch, C.O. Wirsen, and T. Hoaki

During the past decade, a large number of heterotrophic Archaea that grow within a temperature range of approximately 75–110°C have been isolated (Stetter et al. 1990). The nutritional requirements of these organisms are still enigmatic as most of them can only be grown on complex organic media (Jannasch et al. 1992). The following protocol concerns organisms isolated from samples retrieved from hydrothermal vents at the East Pacific Rise (depth 2500–2700 meters), Guaymas Basin (Gulf of California, depth 2000 meters), and Mid-Atlantic Ridge 23°N and 26°N (depth 3600–3700 meters) with the aid of the research submersible ALVIN.

Types of Samples

ANOXIC HOT WATER

Highly reduced hydrothermal fluid is collected with the Edmond-Walden sampler, the intake tube of which is inserted into a vent orifice. On hydraulic triggering, a flushing cycle removes all seawater from the intake tube before the sample of vent water (~800 ml) is drawn in by spring operation. Two samplers can be connected to one intake for simultaneous filling, so that samples can be stored and used separately, for example, for biological and chemical analyses. The sampler is constructed of titanium and tolerates water temperatures up to 400°C. Although it is not usually sterilized, washing the sampler prior to assembly is thought to remove possible contamination by hyperthermophilic Archaea from samples taken earlier. If the sampler is used to sample hydrothermal fluid for the isolation of hyperthermophiles, contamination by other deep-sea bacteria is assumed to be inconsequential.

For enrichment cultures or storage of subsamples, portions are withdrawn from the Edmond-Walden sampler and transferred to Hungate tubes (Bellco Glass 2047-1625) under flushing with oxygen-free N_2 or a mixture of H_2/CO_2 (commonly 4:1). Gassed syringes are used to inject subsamples into medium-filled Hungate tubes. Nonsterile samplers (Go-flow and Niskin-type, General Oceanics 1080 and 1010-5.0) are also used for anoxic cooler waters from "warm" vents where a high H_2S content assures anoxic conditions during the retrieval time. Again, these samplers cannot be sterilized. Niskin-type water samplers with sterilizable plastic bags are used for collecting cold bottom water from the vent surroundings.

HOT SEDIMENTS

Hot sediments are found only at the Guaymas Basin vent site and at shallow marine vent sites such as Lucrino near Naples or Porto di Levante on the island of Vulcano in the Mediterranean Sea where the first marine hyperthermophiles were isolated. Sediments are taken by push corers from the research submersible ALVIN. Small core sections (1–5 cc) are transferred into sample jars containing artificial seawater (ASW, see Appendix 2) reduced with 200–400 μM Na_2S for storage or directly into tubes filled with enrichment media. Most probable numbers (MPNs) can be obtained from quantified sediment suspensions used for the inoculation of a series of tubes filled with growth media.

SOLID POLYMETAL SULFIDES

Pieces of sulfide-covered lava, sulfide rocks, or smoker chimney walls are collected in an aluminum cylinder fastened to ALVIN's basket. This relatively simple container can be closed with the simultaneous release of a reducing agent such as sodium dithionite (final concentration ~0.05%) from a plastic bag. If larger pieces are retrieved unprotected through the oxygenated seawater column, subsamples from the inside may still be reduced enough to contain viable hyperthermophiles. Under gassing (see above), small pieces of the sulfide deposits are transferred into medium-filled culture tubes. Polychaete tube casings collected from black smoker chimney walls were also sources of successful isolations (Jannasch et al. 1988).

Media and Solutions

Marine broth 2216 (Difco 0791-01) is diluted to half strength with artificial seawater (ASW, see Appendix 2, Medium 1) and supplemented with 1% (w/v) elemental sulfur, 20 μM PIPES buffer, and 0.2 ml of 0.2% resazurin. The sulfur is steam-sterilized for 20 minutes at 110°C on three successive days and added to Hungate tubes before the liquid medium is dispensed. The autoclaved medium is generally filter-sterilized (0.2 μm pore size) to remove fine precipitates. The medium is then pipetted into tubes, the gas space flushed with oxygen-free N_2 for 6–10 minutes, and the liquid phase reduced by adding sodium sulfide from a stock solution to a final concentration of 200–400 μM. Nitrogen is rendered oxygen-free by passing it over a heated copper bed. Bis-Tris Propane (Sigma B 6755) (8 μM) can be substituted as a buffer in place of PIPES to prevent the formation of precipitate during autoclaving. Instead of autoclaving, the medium may also be filter-sterilized (0.2-μm pore size) and reduced with sterile sodium sulfide. In certain cases, elemental sulfur can be replaced by cysteine (10 mM) and the medium reduced with sulfide as noted above.

Other organic substrates such as yeast extract, peptone, acetate, maltose, or starch are used at 0.2% together with 1% (w/v) elemental sulfur, 5 ml of the trace minerals stock solution, and 5.0 ml of the vitamin mixture. See Appendix 2 Medium 1 for ASW, Wolf's trace minerals, vitamin mixture, and modified TYEG medium.

Sample Transportation, Storage, and Incubation

Transportation and shipping of sample material (as well as cultures) are preferably done at refrigeration temperatures (3–8°C). Higher temperatures (9–40°C, clearly below growth-range temperatures, the highest possibly occurring during prolonged storage in tropical climates) may encourage profuse growth of mesophilic bacteria, thereby decreasing the relative population size of hyperthermophiles in the sample. Frozen samples are important for certain chemical and biochemical analyses but do not seem to offer an advantage for the preservation of viable Archaea. Larger amounts of anoxic sulfidic sediment samples in jars are less susceptible to oxygenation and can be used for transfer into Hungate tubes in the laboratory. Sample pieces of mineral sulfides can be stored and shipped in seawater reduced with sodium dithonite (0.05%) or sodium sulfide (0.5–1.0 mM). Such pieces can also be stored in gassed Hungate tubes. Water samples keep well in completely filled Hungate tubes. In general, hyperthermophilic Archaea can often be isolated from samples years after storage (Jannasch et al. 1992).

Inoculated media in gas-tight Hungate tubes with screw-cap (Bellco Glass 2047) or aluminum seal closures (Bellco Glass 2048) are incubated, aboard ship or in the home laboratory, in covered water baths at temperatures near 100°C. Convection and gravity ovens can also be used, particularly at temperatures above 100°C. The type of tube used permits pressurization of up to 3 atm, which prevents boiling of the media at <120°C. When growth is indicated by turbidity, subsamples are transferred into fresh media using the Hungate technique and gassed syringes. Transfers are best done at room temperature. Transfers at high temperatures are more susceptible to oxygen toxicity.

Isolation and Purification

Hungate procedures (Balch et al. 1979) are generally applied by gassing with oxygen-free N_2 or the above-mentioned H_2/CO_2 mixture. An anoxic hood is useful for the dissection of smoker wall pieces or sediment cores. The isolation of an organism that dominates an enrichment culture is best carried out by dilution to extinction in a series of tubes with fresh medium. Alternatively, streak cell suspensions on agar plates (3%) containing colloidal sulfur (Wieringa 1966) and the chosen organic substrate. The plates are incubated in a Torbal jar (BBL 60627) flushed with oxygen-free N_2 and incubated under an N_2 atmosphere. Anoxic conditions can also be provided by gas packs (BBL 71040) that produce a CO_2/H_2 atmosphere. These agar plates can be incubated up to 75–80°C. A small dish of Drierite placed at the bottom of the jar absorbs excessive moisture. Colonies are restreaked under the same conditions. For incubation and streaking on plates at higher temperatures, the gelling agent Gelrite (Merck 46-3070-00) is suitable as originally described by Deming and Baross (1986). Colonies can also be grown in and removed under gassing from Gelrite tubes. Pure cultures are grown in larger-volume vessels (e.g., 2.5-liter polycarbonate Fernbach flasks; Nalge 41052800) in a boiling water bath for the production of biomass sufficient, for example, for DNA extraction, enzyme studies, membrane lipid analysis, and fermentation products.

Table 1 Amino Acid Composition of Yeast Extract, Peptone, and Casamino Acids[a]

	Yeast extract (nmole/ml)		Peptone (nmole/ml)		Casamino acids (nmole/ml)
Amino acid	untreated	hydrolyzed	untreated	hydrolyzed	untreated
Asx[b]	16.5	33.6	2.3	27.8	77.5
Thr	11.4	15.8	2.5	14.7	43.6
Ser	21.8	18.0	4.0	28.0	66.6
Glx[c]	48.2	57.4	5.0	58.9	200.1
Pro	9.4	13.9	1.6	89.0	132.9
Gly	16.9	26.4	7.6	251.3	35.4
Ala	50.6	43.1	9.9	61.8	81.5
Cys-Cys	3.5	n.d.	n.d.	n.d.	n.d.
Val	22.9	24.2	5.4	19.6	59.3
Met	5.4	5.2	1.4	5.7	20.1
Ile	17.2	18.2	3.2	12.1	33.4
Leu	30.1	25.9	10.4	23.0	76.5
Tyr	4.1	6.4	3.4	5.2	4.8
Phe	13.4	12.3	6.2	11.7	26.5
His	6.3	5.9	1.9	4.2	24.0
Lys	17.9	24.9	6.1	23.3	77.1
Trp	2.9	n.d.	1.5	n.d.	n.d.
Arg	6.4	9.6	12.3	33.3	26.8

n.d. indicates not detected.

[a] 0.01% each, all Difco. Data obtained with an amino acid analyzer. Hydrolyzation conditions: 6 N HCl at 110°C for 24 hours.

[b] Asx = Asp and Asn combined.

[c] Glx = Glu and Gln combined.

Amino Acid Requirements

The need for complex proteinaceous substrates by the heterotrophic hyperthermophilic Archaea is not well understood. Most of the known isolates grow on yeast extract or peptone but not on casamino acids (Difco 0230-02-0). Table 1 lists the free and the hydrolyzable amino acids in yeast extract and peptone and the composition of casamino acids. Since acid-labile amino acids are lost during hydrolysis of caseine, the source of casamino acids, the need for specific essential amino acids in two archaeal hyperthermophiles, *Desulfuromonas* strain SY and *Pyrococcus* strain GB-D, was revealed (Hoaki et al. 1993).

DETERMINATION OF ESSENTIAL AMINO ACIDS

ASW containing Wolfe's trace minerals (see Appendix 2, Medium 1) and 0.2% resazurin is set at pH 7.0–7.2. A stock amino acid solution is prepared containing the following 20 amino acids at 1.0 g/liter each: glycine, alanine, serine, threonine, cysteine, asparagine, glutamine, leucine, isoleucine, valine, methionine, phenylalanine, tyrosine, tryptophan, proline, aspartic acid, glutamic acid, histidine, lysine, and arginine. An additional 20 solutions are prepared and each solution lacks one of the above amino acids. ASW and the amino acid solutions are autoclaved separately. After cooling, the ASW is supplemented with 5 ml/liter of the filter-sterilized vitamin mixture (see Appendix 2, Medium 1).

Approximately 0.1 g of steam-sterilized sulfur (see above) is added to sterile Hungate tubes. While being flushed with nitrogen gas, the tubes are filled with 9.0 ml of ASW and 1.0 ml of each of the amino acid mixtures (final concentration of each amino acid is 100 mg/liter). The tubes are sealed with butyl rubber stoppers. To reduce the medium, use a 22-gauge needle to inject 0.1 ml of 1% sodium sulfide into each tube, which turns the resazurin colorless. The tubes are then inoculated with cells from the late exponential growth phase of a culture (cooled to room temperature) to a density of about 10^6 cells per milliliter and incubated at the optimum growth temperature of the organism. The incubation time resulting in visible turbidity does vary. Growth is determined by the increase of cell density. Cells are counted using epifluorescence microscopy after staining with acridine orange. To account for "carry over" of an essential amino acid through the inoculum, the growth experiment is repeated by transfer into fresh medium at least once.

COMMENTARY

Since the heterotrophic Archaea under discussion have been isolated from considerable ocean depths (2000–3600 meters), the question of their barophilism (optimum growth at pressures above 1 atm) arises. No distinct pressure adaptation, as observed in other deep-sea isolates (largely aerobic bacteria being collected at much greater depths; see, e.g., Yayanos and DeLong 1987), has been found so far in our isolates (Jannasch et al. 1988, 1992). Furthermore, although increased hydrostatic pressure does increase the growth rate slightly (up to 10%) at the optimum growth temperature, the latter and the maximum growth temperature are not affected. During the same studies, we observed that these strictly anaerobic Archaea are somewhat insensitive to oxygen below the temperature range for their growth. We therefore assume that they remain viable in cold oxygenated water above the sea floor and are thereby able to spread to newly forming vent fields.

REFERENCES

Balch, W.E., G.E. Fox, L.J. Magrum, C.R. Woese, and R.S. Wolfe. 1979. Methanogens: Reevaluation of a unique biological group. *Microbiol. Rev.* **43:** 260–296.

Deming, J.W. and J.A. Baross. 1986. Solid medium for culturing black smoker bacteria at temperatures to 120°C. *Appl. Environ. Microbiol.* **51:** 238–243.

Hoaki, T., C.O. Wirsen, S. Hanzawa, T. Maruyama, and H.W. Jannasch. 1993. Amino acid requirements of two hyperthermophilic archaeal isolates from deep sea vents: *Desulfurococcus* Strain SY and *Pyrococcus* Strain GB-D. *Appl. Environ. Microbiol.* **59:** 610–613.

Jannasch, H.W., C.O. Wirsen, S.J. Molyneaux, and T.A. Langworthy. 1988. Extremely thermophilic fermentative archaebacteria of the genus *Desulfurococcus* from deep-sea hydrothermal vents. *Appl. Environ. Microbiol.* **54:** 1203–1209.

——. 1992. Comparative physiological studies on hyperthermophilic archaea isolated from deep-sea hot vents with emphasis on *Pyrococcus* Strain GB-D. *Appl. Environ. Microbiol.* **58:** 3472–3481.

Stetter, K.O., G. Fiala, G. Huber, R. Huber, and A. Segerer. 1990. Hyperthermophilic microorganisms. *FEMS Microbiol. Rev.* **75:** 117–124.

Wieringa, K.T. 1966. Solid media with elemental sulfur for detection of sulfur-oxidizing microbes. *Antonie v. Leeuwenhoek* **32:** 183–186.

Yayanos, A.A. and E.F. DeLong. 1987. Deep-sea bacterial fitness to environmental temperatures and pressures. In *Current perspectives in high pressure biology* (ed. H.W. Jannasch et al.), pp. 17–32. Academic Press, London.

PROTOCOL 2

Isolation, Growth, and Maintenance of Hyperthermophiles

J.A. Baross

The culture conditions for different species of hyperthermophiles generally reflect the physical and geochemical conditions of their habitat. Hyperthermophiles are defined as microorganisms capable of growth above 90°C and generally have minimum growth temperatures between 65°C and 70°C. Hyperthermophilic species are found in the Archaea and Bacteria domains (Woese et al. 1990) (see Appendix 1). For the purpose of this protocol, hyperthermophiles have been grouped into three categories: (1) neutrophiles, which grow at neutral to slightly acidic conditions; (2) acidophiles, which grow optimally at pH 3 or lower; and (3) (neutrophilic) methanogens (Stetter et al. 1990; Baross 1993). This is a practical grouping and has little to do with nutritional characteristics or phylogeny (a nutritional delineation of hyperthermophiles has little to do with phylogenetic relatedness). Rather, it reflects the acidity of different continental and marine thermal environments, which translates directly into the properties of isolation and culturing procedures and media.

Sampling Methods

Sampling methods differ according to the specific environment of interest. Reports that describe the isolation of hyperthermophiles from terrestrial or shallow marine sulfotaric environments rarely include a detailed description of the sampling protocol or the container used for collection. These descriptions include collecting water or sediment into sterile jars or sterile plastic syringes to sample terrestrial hot springs. Plastic syringes with the top cut off at the needle end are used as coring devices for hot sediments. Shallow marine solfotara sites have been sampled by either scuba divers or surface boat hydrocasts using different coring devices and grab samplers.

Thermal sites associated with submarine hydrothermal vents include hot water, solids from sulfidic smokers and flanges, sediments, and animals; all must be sampled by a submersible. The 755-ml titanium syringe described by Von Damm et al. (1985) is used frequently to obtain hot water samples from smokers and flange pools. Under ideal conditions, the syringe can collect pure hydrothermal fluids. In general, however, there is some degree of mixing with ambient seawater so that confirming the source of thermophilic bacteria subsequently isolated or detected is rarely a straightforward task. An improved sampling manifold syringe system called the "submersible-coupled in situ sensing and sampling system" (SIS^3) (for details, see Massoth et al. 1989) accommodates three titanium syringes and offers the advantages of flushing the syringes with hot water before sampling and monitoring the water temperature both at the source of the sample and at the intake of the sampling syringe. We have had considerable success with the SIS^3 in obtaining hydrothermal fluids minimally compromised by seawater (Straube et al. 1990). Smoker sulfides and flange solids are usually sampled by breaking apart larger structures with the submersible arm. Solid samples as well as animal specimens should be placed in an insulated "live box" on the submersible basket so as to lessen the chance of losing the sample upon surfacing and to help prevent the sample from reaching warm surface temperatures.

Since most of the known hyperthermophiles are anaerobic, it is advisable to inoculate samples directly into anaerobic Balch tubes (Bellco Glass 2048) capped with gas-impermeable black butyl septum-type rubber stoppers (Bellco Glass 2058-11800) secured with aluminum crimp seals (Bellco Glass 2048-11020) or to store samples in BBL anaerobic jars until they can be inoculated into suitable media. An anaerobic jar with a BBL Gas-Pac (70304; H_2/CO_2) is convenient for creating anaerobic conditions in the field. Huber et al. (1991) added appropriate concentrations of sodium dithionate and sodium sulfide to sample bottles until they became anaerobic using resazurin as the indicator. In general, however, most strains of anaerobic hyperthermophiles are stable for at least a short time in the presence of oxygen as long as they are held at temperatures below their minimum for growth. It is recommended that an anaerobic transfer hood be used for initial enrichments for hyperthermophiles from solid samples such as sediments, sulfides, and vent animals, and when transferring cultures inoculated in the field. Inexpensive, disposable plastic glove bags (Instruments for Research and Industry R-27-17-H) are suitable for shipboard use. Use either nitrogen or argon gases with these glove bags.

In extremely acidic environments ($pH < 3$), only species of *Sulfolobus* and *Acidianus* will likely be isolated. Special care must be taken to adjust the pH of the sample to 5.5–6.0 until the sample is prepared for culturing. *Sulfolobus* spp. can grow over a pH range of 1 to 4; while growing, they maintain a higher internal pH. The internal pH will quickly reach surrounding values if the cells are not growing, which will result in cell lysis. So far, no strains of obligately barophilic or halophilic species of hyperthermophiles have been isolated.

Isolation and Maintenance of Hyperthermophiles

Most hyperthermophiles are difficult or impossible to cultivate on solid media, and it is thus necessary to use a dilution-to-extinction series to obtain pure cultures. A satisfactory procedure involves 1:3 dilutions of a culture in the late log phase of growth, starting the dilution series at three orders of magnitude below the actual number of cells in the culture to be diluted as determined by epifluorescence microscopy (Porter and Feig 1980). The dilutions should be carried out to two orders of magnitude higher than the estimated number of cells in the culture. Repeat the dilution-to-extinction series at least three times using the highest dilution tubes showing growth. In some cases where there is still evidence of a mixed culture, each dilution in the dilution-to-extinction series should be run in triplicate.

Appendix 2 (Medium 2) includes recipes for synthetic seawater, *Sulfolobus* medium, Trace elements solutions A and B, Vitamin solution, and *Methanothermus* mineral solutions 1 and 2, which can be used for the growth and isolation of most known species of hyperthermophiles. The following sections on neutrophiles, acidophiles, and methanogens describe specific formulations and procedures for making media. It should be kept in mind that other species of hyperthermophiles not yet isolated may have unusual nutrient, electron acceptor, trace metal, and organic compound requirements. Medium formulations must reflect the physical and geochemical properties of specific environments.

Specific Growth Media and Preparation

Neutrophiles (Nonmethanogenic)

Most of the neutrophilic, anaerobic, and heterotrophic hyperthermophiles can grow on media containing various hydrolyzed protein preparations, such as peptones and trypticase soy, and yeast-extract-supplemented trace minerals and elemental sulfur. In general, marine isolates have a seawater requirement. The most commonly used medium formulation for isolation of both sulfur- and nonsulfur-dependent marine heterotrophs consists of synthetic seawater (see Appendix 2), a source of complex organic carbon and sulfur. The following is a useful medium formulation for initial isolation of a variety of hyperthermophiles from diverse marine environments (per liter of synthetic seawater):

Trace elements A (see Appendix 2)	10 ml
Trace elements B (see Appendix 2)	1 ml
$(NH_4)_2SO_4$ and KH_2PO_4 solution (per liter)	10 ml
43.0 g $(NH_4)_2SO_4$	
3.6 g KH_2PO_4	
FeEDTA solution (per liter)	2 ml
1.54 g $FeSO_4{\cdot}7H_2O$	
2.06 g Na_2EDTA	
Yeast extract (Difco 0127-01-7)	1 g
Bacto-peptone (Difco 0118-01-8)	1 g
0.1% Resazurin solution (0.001 g final concentration)	1 ml

Adjust pH to 6.5. Filter-sterilize medium and transfer to Balch tubes (18 x 150 mm; Bellco Glass 2047) or Wheaton "400" clear serum bottles (VWR Scientific 16171-341). When appropriate, add 5 g/liter sterile elemental sulfur (S^o) to each tube or bottle (sterilize S^o separately by steaming for 1 hour on three successive days at 100°C). Cap the bottles or tubes with gas-impermeable black butyl septum-type stoppers (Bellco Glass 2048-11800) secured with aluminum crimp seals (Markson Science 026 series or Bellco Glass 2048-11020) using a crimper (Bellco Glass 2048-10020).

Replace air with ultra high-purity-grade argon or a blend of H_2/CO_2 if appropriate by alternately pulling a vacuum and saturating with gas using a degassing manifold. After degassing, remove any traces of oxygen by adding sterile 3% Na_2S solution to a final concentration of 0.03%. Apply a headspace pressure of approximately 100 kPa to all tubes to prevent air intrusion.

This medium has been used successfully to grow marine *Pyrococcus* spp., *Thermococcus* spp., *Staphylothermus* spp., and related heterotrophic cocci such as ES1 and ES4 (Pledger and Baross 1989, 1991), *Pyrococcus abyssi* (Erauso et al. 1993), and *Pyrodictium brockii.* Many strains of these organisms have been grown in Bacto Marine Broth 2216 (Difco 0791-17-4). This medium contains synthetic sea salts, peptone, and beef extract. Similarly, *Hyperthermus butylicus* had been cultured initially with 0.6% tryptone and has subsequently been found to require peptides for growth (Zillig et al. 1990). *Pyrococcus* strain GB-D isolated from Guaymas hot springs will not grow on purified proteins such as casein, but it will grow on media with peptone, tryptone, and yeast extract (Jannasch et al. 1992). Marine species of *Thermotoga* (Bacteria domain) have been cultured with 0.1% yeast extract and 0.2% glucose. Species of *Archaeoglobus* recently isolated from submarine hydrothermal vent environments reduce SO_4^{2-}, $S_2O_3^{2-}$, or SO_3^{2-} to H_2S and can utilize a variety of carbon sources, including proteins, sugars, lactate, formate, and pyruvate (Stetter 1988; Zillig 1989; Burggraf et al. 1990). One species, *A. fulgidus,* can grow chemolithotrophically only with $S_2O_3^{2-}$ as the electron acceptor and H_2 as the energy source.

Nonsalt requiring heterotrophic species of both Archaea and Bacteria can be cultured in the *Sulfolobus* medium (see Appendix 2) adjusted to pH 5.5–7.0 and supplemented with appropriate carbon sources and electron acceptors. The sulfur-requiring *Thermoproteus* species will grow heterotrophically on a variety of carbon sources, including glucose, starch, and amino acids, or chemolithotrophically by using S^o and H_2 as energy sources and CO_2 as the carbon source (Zillig 1989). *Thermoproteus* species are slightly acidophilic; the pH of the medium should be adjusted to between 5 and 6.5 depending on the species.

Most of the heterotrophic Archaea have minimum growth temperatures between 60°C and 70°C and maximum growth temperatures to 110°C. Choosing incubation temperatures will depend on the species. However, for the initial isolation of hyperthermophilic Archaea, 85–90°C incubation is recommended in order to reduce the chances of mixed cultures of organisms that grow in the 80°C range. This is particularly true with submarine hydrothermal vent samples in which we have frequently encountered organisms resembling *Clostridium* spp. (bacteria) with maximum growth temperatures in the low 80°C range.

Few published procedures exist for culturing heterotrophic hyperthermophiles on solid medium (for review, see Wiegel 1986). In general, most solidifying agents are too unstable at temperatures much above 70°C for extended periods. However, there are reports of some species of hyperthermophilic heterotrophs, such as *Pyrococcus furiosus,* that form colonies on agar media at 70°C (Fiala and Stetter 1989). In my experience, none of the heterotrophic hyperthermophiles isolated from submarine hydrothermal vent environments were found to form colonies on agar plates. This may be because some species cannot form colonies on surfaces or because 70°C is too close to their minimum growth temperature. Agar media evaporate extremely rapidly at temperatures above 70°C. Some success has been achieved using the gellan gum, Gelrite (Kelco Division of Merck GR79008A [lot no.]), in shake tubes with environmental samples incubated at 90–120°C (Deming and Baross 1986) and as a plating medium at 99°C for the isolation of pure cultures of *Hyperthermus butylicus* (Zillig et al. 1990). Schleper et al. (1992) (see also Protocol 3) reported the use of Gelrite in a plaque assay medium for enumerating the "virus-like" particle SSV1 from *Sulfolobus shibatae* on lawns of *S. solfataricus*. There are few published reports of hyperthermophilic Archaea forming colonies on solid media. Stetter (1989) indicates that *Methanothermus fervidus* can be cultured at 85°C on plated medium solidified by polysilicate and incubated with H_2 and CO_2 in a pressure cylinder (Balch and Wolfe 1976). *M. fervidus* will not grow on agar (Stetter 1989).

The preparation of solid media with Gelrite gum requires some practice since it solidifies rapidly when divalent cations such as calcium and magnesium are added. When preparing media with Gelrite, it is important to omit the calcium salts, and for seawater medium the magnesium salts as well, from the mineral medium to be autoclaved. After autoclaving, the melted Gelrite solution (8 g of gellan gum per liter; however, the concentration should be determined empirically with different batches of Gelrite) should be transferrred to an oil bath at approximately 100°C. Additional sterile ingredients preheated to 100°C can be added to the Gelrite. The calcium salts and trace elements should be added last just prior to distribution of medium into shake tubes containing environmental samples (Deming and Baross 1986) or into petri plates (Wiegel 1986; Zillig et al. 1990). The Gelrite plating medium described by Zillig contained colloidal sulfur for culturing sulfur-dependent hyperthermophiles. The Gelrite was dissolved in boiling water and solidified by addition of $CaSO_4$ (1 g/liter), poured into petri plates, allowed to solidify, and then soaked with 5 ml of a solution containing saturated levels of elemental sulfur in 1 M $(NH_4)_2S$ for 2 minutes. After rinsing with water, the colloidal sulfur was precipitated by addition of 5–10 ml of 1 M H_2SO_4 for 2 minutes and then washed thoroughly with water. The plates were soaked with liquid culture medium overnight, dried at 37°C, and then stored in an anaerobic chamber. Dilutions of liquid cultures were plated directly onto these plates and incubated at 99°C. For additional procedures on the isolation of plateable hyperthermophiles using Gelrite media, see Protocol 3.

Acidophiles

The thermoacidophilic Archaea include two genera in the family Sulfolobaceae, *Sulfolobus* and *Acidianus*. Known species include strict and facultative aerobes and obligate and facultative chemolithotrophs. All grow optimally at a pH of 2–3. *Sulfolobus* spp. are enriched in *Sulfolobus* medium (see Appendix 2) and incubated aerobically at 70–80°C; they have also been isolated in the field by adding 0.1% yeast extract (using a 1% solution) to thermal spring water and acidifying to pH 2. Both *Sulfolobus* and the related genus *Acidianus* will grow chemolithotrophically with S^0 and O_2. *A. infernus* is an obligate chemolithotroph capable of growing aerobically or anaerobically. Anaerobic growth involves the reduction of S^0 by H_2 using an H_2/CO_2 gas mixture. Similarly, *Sulfolobus metallicus* is a strictly chemolithotrophic, obligately aerobic organism capable of mobilizing metals from ores (Huber and Stetter 1991). This species can oxidize S^o, pyrite, chalcopyrite, and sphalerite.

There are reports of *Sulfolobus* and *Acidianus* forming colonies aerobically and anaerobically on medium solidified by 10% starch and on Gelrite (Zillig 1989; Schleper et al. 1992). Other solidifying agents would be too unstable at the required acidic pH.

Methanogens

Most of the species of hyperthermophilic methanogens thus far described belong to three genera. All of the marine species belong to the genera *Methanococcus* and *Methanopyrus*, whereas *Methanothermus* spp. have been isolated only from terrestrial solfatara environments. Other thermophilic methanogens, such as *Methanobacterium thermoautotrophicum*, grow at temperatures above 55°C but generally not above 70°C. All of the hyperthermophilic methanogens can grow with H_2 and CO_2, but the growth of many species is stimulated by addition of complex organic material such as yeast extract, trypticase, and peptone or requires vitamins. Some species of *Methanococcus* will grow on formate.

Marine Methanogens

A useful medium for initial isolation of marine hyperthermophilic methanogens includes (per liter synthetic seawater):

Trace elements A (see Appendix 2)	10 ml
Trace elements B (see Appendix 2)	1 ml
$(NH_4)_2SO_4$ and KH_2PO_4 solution (per liter)	10 ml
43.0 g $(NH_4)_2SO_4$	
3.6 g KH_2PO_4	
Vitamin solution (see Appendix 2)	10 ml
10% $NaHCO_3$ solution	10 ml
0.1% Resazurin solution (0.001 g final concentration)	1 ml

Adjust pH to 7.0. Filter-sterilize medium and transfer to Balch tubes (18 x 150 mm; Bellco Glass 2047) or Wheaton "400" clear serum bottles (VWR Scientific 16171-341). Cap the bottles or tubes with gas-impermeable, black septum-type stoppers (Bellco Glass 2048-11800). Remove air as described above for neutrophiles and replace with H_2/CO_2 leaving a pressure in the tubes or bottles of 200 KPa. After degassing, remove any traces of oxygen by adding sterile 3% Na_2S solution to a final concentration of 0.03%.

Other reducing agents that have been used with methanogens and other anaerobes include cysteine-sulfide, titanium citrate, dithiothreitol, and thioglycollate; they are generally used at a final concentration of 0.025% to 0.05%. In general, it is recommended that various carbon sources be added to the medium for initial isolation of methanogens. Usually, 2 g of both yeast extract and trypticase are added per liter to the medium. Some species of hyperthermophilic *Methanococcus* will grow in a medium containing 2.5–5 g/liter of formate as a carbon source. Selenium stimulates growth of most species of *Methanococcus*.

Nonmarine Methanogens

The recommended medium for isolation of hyperthermophilic nonmarine species of *Methanothermus* is described by Balch et al. (1979) and includes (per 905 ml of distilled H_2O): 37.5 ml each of *Methanothermus* Mineral Solutions 1 and 2 (see Appendix 2, Medium 2).

Trace elements A (see Appendix 2)	10 ml
Trace elements B (see Appendix 2)	1 ml
Vitamin solution (see Appendix 2)	10 ml
Yeast extract (Difco)	2 g
BBL-Trypticase	2 g
Sodium acetate	5 g
$NiCl_2 \cdot 6H_2O$	1.6 mg
$FeSO_4 \cdot 7H_2O$	2 mg
Coenzyme M	1.6 μg
10% $NaHCO_3$ solution	6 g
Cysteine hydrochloride	0.25 g
3% $Na_2S \cdot 9H_2O$ solution	0.25 g
0.1% Resazurine solution (0.001 g/liter final concentration)	1 ml

Adjust pH to 6.7 with acetic acid. Medium preparation and sterilization are the same as those described for marine methanogens. For additional media formulations and procedures for isolation and maintenance of methanogens, see Balch et al. (1979).

Incubation

Special equipment is not necessary for routine incubation of hyperthermophiles up to temperatures of 110°C and pressures to 500 kPa (pressure achieved in sealed Bellco tubes). Tubes or serum bottles can be incubated in commercial water baths filled with oil. Dimethyl/silicone oil (Thomas Scientific Silicone Fluid SF 96/50) and related oils are stable to 300°C and are recommended over organic oils or solvents for their thermal stability. Bellco tube cultures can also be incubated in heating blocks (Reacti-Therm heating modules 18800, series 343) filled with oil (Pledger and Baross 1989). Air incubators can be used for culturing hyperthermophiles at temperatures below 85°C, but the temperature will fluctuate considerably if the cultures must be accessed frequently. Standard water baths filled with sand are useful for isolation of hyperthermophiles on shipboard. It is important to note temperature gradients established vertically and horizontally in the sand bath. The temperature must be measured at the location and depth in the sand where culture tubes are placed.

Some progress has been made toward the growth of hyperthermophiles in continuous culture to obtain high cell biomass for biochemical and molecular biology studies. Brown and Kelly (1989) describe an inexpensive and easily constructed reactor for the continuous culture of hyperthermophilic heterotrophs at temperatures to 100°C (see Protocol 4). All of the components for this reactor can be purchased "off the shelf."

COMMENTARY

- Some species of hyperthermophiles grow optimally with doubling times of less than 1 hour and reach densities of greater than 10^9 cells per milliliter.

Other species grow extremely slowly (>10-hour doubling times) and will only reach densities of 1–5 x 10^7 cells per milliliter. This is particularly true for some species of *Thermoproteus* cultured either heterotrophically or chemolithotrophically. It is advisable to use microscopic methods to assess growth in enrichment cultures in instances where there is no evidence of visible turbidity.

- Many species of hyperthermophilic heterotrophs will lyse relatively rapidly when left at high growth temperatures beyond the early stationary growth phase. Most isolates will remain stable for weeks at room or refrigerated temperatures under anaerobic conditions.

- Mukund and Adams (1991) have reported that *Pyrococcus furiosus* and other species of heterotrophic hyperthermophiles have unique tungsten enzymes and therefore require a higher concentration of tungsten for optimal growth than is normally found in most trace elements mixtures. It is recommended that 10 μM Na_2WO_4 be added per liter of medium for cultivation of most species of heterotrophic, hyperthermophilic Archaea. Other species of hyperthermophiles have different metal and electron acceptor requirements; it is important to consult reports on individual species for specific nutrient requirements.

REFERENCES

Balch, W.E. and R.S. Wolfe. 1976. New approach to the cultivation of methanogenic bacteria: 2-Mercaptoethane-sulfonic acid (HS-CoM)-dependent growth of *Methanobacterium ruminantium* in a pressurized atmosphere. *Appl. Environ. Microbiol.* **32:** 781–791.

Balch, W.E., E.E. Fox, L.J. Magrum, C.R. Woese, and R.S. Wolfe. 1979. Methanogens: A reevaluation of a unique biological group. *Microbiol. Rev.* **43:** 260–296.

Baross. J.A. 1993. Isolation and cultivation of hyperthermophilic bacteria from marine and freshwater habitats. In *Current methods in aquatic microbial ecology* (ed. P.F. Kemp et al.), pp. 21–30. Lewis Publishers, Boca Raton, Florida.

Brock, T.D., K.M. Brock, R.T. Belly, and R.L. Weiss. 1972. *Sulfolobus:* A new genus of sulfur-oxidizing bacteria living at low pH and high temperature. *Arch. Microbiol.* **84:** 54–68.

Brown, S.H. and R.M. Kelly. 1989. Cultivation techniques for hyperthermophilic archaebacteria: Continuous culture of *Pyrococcus furiosus* at temperatures near 100°C. *Appl. Environ. Microbiol.* **55:** 2086–2088.

Burggraf, S., H.W. Jannasch, B. Nicolaus, and K.O. Stetter. 1990. *Archaeoglobus profundus* sp. nov. represents a new species within the sulfate-reducing archaebacteria. *Syst. Appl. Microbiol.* **13:** 24–28.

Burggraf, S., G.J. Olsen, K.O. Stetter, and C.R. Woese. 1992. A phylogenetic analysis of *Aquifex pyrophilus. Syst. Appl. Microbiol.* **15:** 352–356.

Deming, J.W. and J.A. Baross. 1986. Solid medium forculturing black smoker bacteria at temperatures to 120°C. *Appl. Environ. Microbiol.* **51:** 238–243.

Erauso, G., A.L. Reysenbach, A. Godfroy, J.-R. Meunier, B. Crump, F. Partensky, J.A. Baross, V.T. Marteinsson, N.R. Pace, G. Barbier, and D. Prieur. 1993. *Pyrococcus abyssi* sp. no., a new hyperthermophilic archaeon isolated from a deep-sea hydrothermal vent. *Arch. Microbiol.* **160:** 338–349.

Fiala, G. and K.O. Stetter. 1989. *Pyrococcus.* In *Bergey's manual of systematic bacteriology* (ed. J.T. Staley et al.), vol. 3, pp. 2237–2240. Williams & Wilkins, Baltimore.

Huber, G. and K.O. Stetter. 1991. *Sulfolobus metallicus,* sp. nov., a novel strictly chemolithoautotrophic thermophilic archaeal species of metal-mobilizers. *Syst. Appl. Microbiol.* **14:** 372–378.

Huber, G., R. Huber, B.E. Jones, G. Lauerer, A. Neuner, A. Segerer, K.O. Stetter, and E.T. Degens. 1991. Hyperthermophilic archaea and bacteria occurring within Indonesian hydrothermal areas. *System. Appl. Microbiol.* **14:** 397–404.

Huber, R., T. Wilharm, D. Huber, A. Trincone, S. Burggraf, H. König, R. Rachel, I. Rockinger, H. Fricke, and K.O. Stetter. 1992. *Aquifex pyrophilus* gen. nov. sp. nov. represents a novel group of marine hyperthermophilic hydrogen-oxidizing bacteria. *Syst. Appl. Microbiol.* **15:** 340–351.

Jannasch, H.W., C.O. Wirsen, S.J. Molyneaux, and T.A.

Langworthy. 1992. Comparative physiological studies on hyperthermophilic archaea isolated from deep-sea hot vents with emphasis on *Pyrococcus* strain GB-D. *Appl. Environ. Microbiol.* **58:** 3472–3481.

Massoth, G.J., H.B. Milburn, S.R. Hammond, D.A. Butterfield, R.E. McDuff, and J.E. Lupton. 1989. The geochemistry of submarine venting fluids at Axial Volcano, Juan de Fuca Ridge: New sampling methods and VENTS Program rationale. In *Global venting, midwater, and benthic ecological processes* (ed. M.P. De Luca and I. Babb), pp. 29–59. National Undersea Research Program Res. Rep. 88-4. National Oceanic and Atmospheric Administration, Rockville, Maryland.

Mukund, S. and M.W.W. Adams. 1991. The novel tungsten-iron-sulfur protein of the hyperthermophilic archaebacterium, *Pyrococcus furiosus*, is an aldehyde ferredoxin oxidoreductase. *J. Biol. Chem.* **266:** 14208–14216.

Pledger, R.J. and J.A. Baross. 1989. Characterization of an extremely thermophilic archaebacterium isolated from a black smoker polychaete (*Paralvinella* sp.) at the Juan de Fuca Ridge. *Syst. Appl. Microbiol.* **12:** 249–256.

———. 1991. Preliminary description and nutritional characterization of a chemoorganotrophic archaeobacterium growing at temperatures of up to 110°C isolated from a submarine hydrothermal vent environment. *J. Gen. Microbiol.* **137:** 203–211.

Porter, K.G. and Y.S. Feig. 1980. The use of DAPI for identifying and counting microflora. *Limnol. Oceanogr.* **25:** 943–948.

Schleper, C., K. Kubo, and W. Zillig. 1992. The particle SSV1 from the extremely thermophilic *Sulfolobus* is a virus: Demonstration of infectivity and of transfection with viral DNA. *Proc. Natl. Acad. Sci.* **89:** 7645–7649.

Stetter, K.O. 1988. *Archaeoglobus fulgidus* gen., sp. nov.: A new taxon of extremely thermophilic archaebacteria. *Syst. Appl. Microbiol.* **10:** 172–173.

———. 1989. *Methanothermaceae.* In *Bergey's manual of systematic bacteriology* (ed. J.T. Staley et al.), vol. 3, pp. 2183–2184. Williams & Wilkins, Baltimore.

Stetter, K.O., G. Fiala, G. Huber, and A. Segerer. 1990. Hyperthermophilic microorganisms. *FEMS Microbiol. Rev.* **75:** 117–124.

Straube, W.L., J.W. Deming, C.C. Somerville, R.R. Colwell, and J.A. Baross. 1990. Particulate DNA in smoker fluids: Evidence for the existence of microbial populations in hot hydrothermal systems. *Appl. Environ. Microbiol.* **56:** 1440–1447.

Von Damm, K.L., J.M. Edmond, B.C. Grant, C.I. Measures, B. Walden, and R.F. Weiss. 1985. Chemistry of submarine hydrothermal solutions at 21°N, East Pacific Rise. *Geochim. Cosmochim. Acta* **49:** 2197–2220.

Wiegel, J. 1986. Methods for isolation and study of thermophiles. In *Thermophiles: General, molecular, and applied microbiology* (ed. T.D. Brock), pp. 17–37. John Wiley & Sons, New York.

Woese, C.R., O. Kandler, and M.L. Wheelis. 1990. Towards a natural system of organisms: Proposal for the domains *Archaea, Bacteria,* and *Eucarya. Proc. Natl. Acad. Sci.* **87:** 4576–4579.

Zillig, W. 1989. *Thermoproteales.* In *Bergey's manual of systematic bacteriology* (ed. J.T. Staley et al.), vol. 3, pp. 2240–2246. Williams & Wilkins, Baltimore.

Zillig, W., I. Holz, D. Janekovic, H.P. Klenk, E. Imsel, J. Trent, S. Wunderl, V.H. Forjaz, R. Coutinho, and T. Ferreira. 1990. *Hyperthermus butylicus,* a hyperthermophilic sulfur-reducing archaebacterium that ferments peptides. *J. Bacteriol.* **172:** 3959–3965.

PROTOCOL 3

Plate Cultivation Technique for Strictly Anaerobic, Thermophilic, Sulfur-metabolizing Archaea

G. Erauso, D. Prieur, A. Godfroy, and G. Raguénes

Described here is the preparation of a solid medium that contains finely dispersed elemental sulfur and allows growth of extremely thermophilic, sulfur-utilizing anaerobes. This medium is solidified with Gelrite, a thermostable gelling agent, which has been used previously for plating aerobic thermophiles (Lin and Casida 1984) or isolating colonies of hyperthermophiles in Gelrite shake tubes (Deming and Baross 1986). Although this system was originally used to isolate new hyperthermophiles (Erauso et al. 1993), it has since been utilized in our laboratory to cultivate members of the genera *Thermococcus* and *Pyrococcus*. Applications include strain purification, determination of viable counts, and isolation of mutants for genetic studies.

General techniques for cultivation of anaerobic Archaea used in these protocols are those originally described by Balch and Wolfe (1976). For details, see Protocols 1 and 2.

MATERIALS

Double-strength medium (1 liter) (see Appendix 2, Medium 3)

Reducing agent: $Na_2S{\cdot}9H_2O$ (5%, w/v). Adjust pH to 7.5, filter-sterilize, and degas under oxygen-free N_2.

Culture and dilution medium: Reduce medium (see Appendix 2) with 5 ml of reducing agent per liter. For liquid culture, add 5 g of S^o (steam-sterilized at 110°C for 1 hour for three successive days) per liter of medium.

Polysulfides solution: Dissolve 10 g of $Na_2S{\cdot}9H_2O$ into 15 ml of deionized H_2O and add 3 g of sulfur flowers. Sterilize by filtration using a 0.2-μm pool size filter (PolyLabo 22676) and keep under N_2.

Gasses: $N_2/H_2/CO_2$ (90:5:5) and N_2, analytical grade, passed through an oxygen trap cartridge to remove traces of oxygen (Oxyclear, Altech 8864)

Palladium cold catalyst (Oxoid cold catalyst sachet, Don Whitley Scientific, or palladium catalyst BASF), for use in the anaerobic chamber or in the anaerobic jars.

Gelrite (Merck) or Phytagel (Sigma P 8169)

Anaerobic chamber (La Calhène)

Anaerobic jars (Don Whitley Scientific DNWA005)

Serum bottles (250 ml; OSI A20.705.47)

Stirring bars
Glass petri dishes (8-cm diameter; OSI 14.160.51)
Electric pipettor (PolyLabo 19239)
Glass pipettes (10 ml, sterile; PolyLabo 98735 or equivalent)
Magnetic stirrer (heating) (Bioblock Scientific A94388 or equivalent)
Plastic disposable spreader
Whatman 3M paper
Thermoblock (Thermolyne, Bioblock Scientific A92607)

METHODS

Culturing on Supporting and Overlay Gels

1. To 100 ml of deionized H_2O in 250-ml serum bottles, add 3 g of Gelrite and mix vigorously to dissolve. Prepare double-strength medium, adjust to pH 6.8, and dispense 100 ml into 250-ml serum bottles containing a stirring bar. Sterilize by autoclaving both solutions at the same time for 20 minutes at 120°C.

2. Immediately after autoclaving, transfer the solutions to a 90°C oven so that they remain liquid until use. Use half the bottles for the supporting gel and half the bottles for the overlay.

3. *For the supporting gel:* Slowly pour the Gelrite solution into the double-strength medium while stirring at 85–90°C on a heating magnetic stirrer. This system keeps the medium in liquid form during distribution in plates. Add 2.6 ml of sterile prewarmed (90°C) 0.1 N HCl (final pH 6.8). Quickly distribute 10 ml per petri dish, using an electric pipettor and 10-ml sterile glass pipettes. The medium solidifies very quickly, forming the supporting gel.

4. *For the overlay gel:* Proceed as in step 3, without adding HCl. Add 0.2 ml of polysulfides solution (the medium turns green-yellow since the polysulfides act as a reducing agent). Distribute 10 ml per plate over the supporting gel. Solidification of the overlay gel occurs quickly. The HCl from the bottom layer diffuses through the top layer and precipitates sulfur from polysulfides as a very fine suspension.

Drying and Reduction of Plates

5. Evaporate all moisture formed on the surface of the gel after solidification by placing the plates open in an inverted position in an air-incubator for 1 hour at 40°C.

6. Transfer the plates into the anaerobic chamber. During the vacuum/flush cycling of the airlock, the vacuum should not exceed 30 mbars (beyond this limit, the gel deteriorates).

7. Store the plates in the anaerobic chamber for at least 24 hours to allow oxygen to diffuse from plates.

8. To ensure full reduction before inoculation, place the plates in anaerobic jars (inside the anaerobic chamber) with a small piece of Whatman paper

wetted with 0.5 ml of $Na_2S \cdot 9H_2O$ 5% (w/v) and 0.5 ml of 0.1 N HCl (the mixture of both compounds produces H_2S). Ensure that the plates turn from pink to colorless.

Inoculation of Plates and Incubation

9. Use liquid medium without sulfur to dilute the inocula as desired. From appropriate dilutions, spread 0.1-ml aliquots onto plates using a plastic disposable spreader.

10. Invert plates in an anaerobic jar with a layer of silica gel to eliminate excessive moisture due to evaporation during incubation at high temperatures. Add a palladium catalyst bag and lock the lid on the jar.

11. Replace gas (if necessary) inside the jar with appropriate atmosphere using a gassing manifold. Incubate the anaerobic jars in an oven at the desired temperature.

Alternative Protocol

1. Prepare plates with the supporting gel as described in step 3. Transfer supporting gel (step 3) into the anaerobic chamber and proceed as described in steps 7 and 8 above.

2. Prepare double-strength medium and Gelrite solution (1.5 g per 100 ml of water) as described above. Degas the solutions as described in Protocol 2 and autoclave for 20 minutes at 120°C.

3. Immediately after autoclaving, transfer the solutions into the anaerobic chamber and mix as described in step 4 above.

4. Distribute 9 ml of soft gel per preheated tube and keep at 90°C in a thermoblock until use.

5. Dilute inocula in prewarmed medium (90°C) and add 1 ml of the appropriate dilution to the double-strength soft gel. Mix thoroughly and pour on the supporting gel in the plate.

6. Incubate plates under the same conditions described above.

COMMENTARY

- The medium was originally described by Fiala and Stetter (1986). The concentration of Gelrite was optimized on the basis of the initial gel strength (determined empirically). In this respect, we found that decreasing the $MgSO_4 \cdot 7H_2O$ concentration from the original formulation to 1 g/liter did not alter the growth of *Thermococcus* and *Pyrococcus* spp., but it kept the medium liquid at lower temperature and was thus more convenient to handle. Because we used this medium to cultivate marine organisms, we also increased the NaCl concentration from the original formulation to 20 g/liter. However, we recommend that the approximately 350 mM Na^+ not be exceeded because gel strength is compromised beyond this limit, even with increasing the amount of Gelrite. Since Gelrite needs divalent cations (such as Mg^{++} and Ca^{++}) to solidify (Moorhouse et al. 1981), perhaps the Na^+ ion competes for the same sites and inhibits gel polymerization. For organisms that require higher NaCl concentration, such as *Thermococcus celer* (Zillig et al. 1983), we suggest equilibrating the plates with the liquid culture medium. Solid medium can also be prepared using "sea salt" (30 g/liter) from Sigma (S 9883) as mineral basis.

• The basic protocol is more convenient because plate preparation can be performed without special attention to anaerobiosis (the operations are done in a sterile hood). The alternative protocol is more tedious, but it gave better results for some of our recent isolates (G. Erauso, unpubl.). To save time, it is also possible to do all of the steps in the anaerobic chamber. In this case, steps 5 and 7 can be omitted. However, manipulation of large volumes of hot medium in the anaerobic chamber is delicate and can be hazardous.

• The plates obtained following these protocols were stored unaltered for a period of 3 weeks at 100°C (the highest temperature tested).

• For *Thermococcus* and *Pyrococcus* spp., colonies (1–3 mm diameter) appear within 2–5 days of incubation, in the temperature range of 80–95°C, depending on the organisms. Colonies are usually surrounded by a clear halo, indicating sulfur utilization (Fig. 1).

• Colonies can be transferred to liquid medium with high efficiency (~90%) or streaked on fresh plates. It should be noted that while streaking on the surface of the plates, sulfur may crystallize to form large particles within clear areas, and this may complicate the interpretation of results.

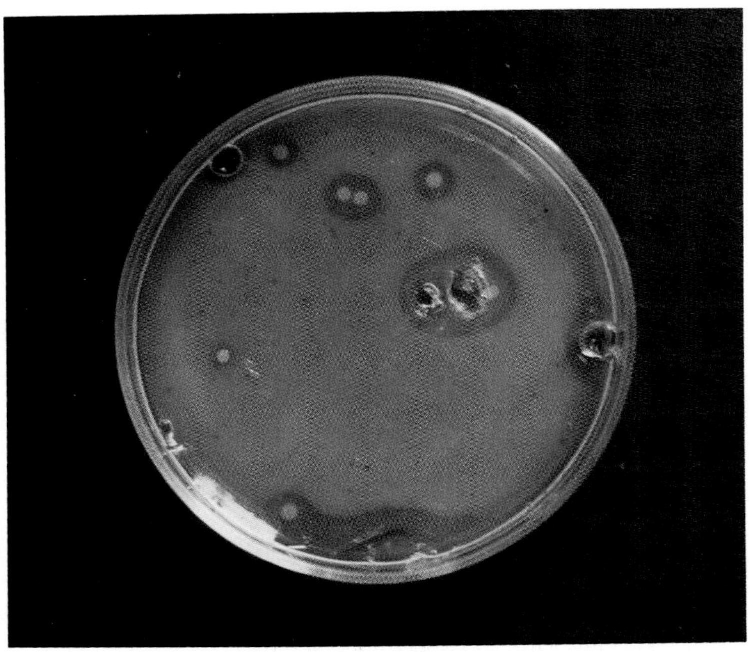

Figure 1 Colonies of *Pyrococcus abyssi* (Erauso et al. 1993) After 3 Days of Incubation at 90°C on a Gelrite Plate (8 cm in diameter). The clear halo surrounding the colonies indicates sulfur utization.

REFERENCES

Balch, W.E. and R.S. Wolfe. 1976. New approach to the cultivation of methanogenic bacteria: 2-Mercapto-ethanesulfonic acid (HS-CoM)-dependent growth of *Methanobacterium ruminantium* in a pressurized atmosphere. *Appl. Environ. Microbiol.* **32:** 781–791.

Balch, W.E., G.E. Fox, L.J. Magrum, C.R. Woese, and R.S. Wolfe. 1979. Methanogens: Reevaluation of a unique biological group. *Microbiol. Rev.* **43:** 260–296.

Deming, J.W. and J.A. Baross. 1986. Solid medium for culturing black smoker bacteria at temperatures to 120°C. *Appl. Environ. Microbiol.* **51:** 238–243.

Erauso, G., A.L. Reysenbach, A. Godfroy, J.R. Meunier, B. B. Crump, F. Partensky, J. Baross, V. Marteinsson, G. Barbier, N.R. Pace, and D. Prieur. 1993. *Pyrococcus abyssi* sp., nov., a new hyperthermophilic archaeon isolated from a deep-sea hydrothermal vent. *Arch. Microbiol.* **160:** 338–349.

Fiala, G. and K.O. Stetter. 1986. *Pyrococcus furiosus* sp. nov. represents a novel genus of marine heterotrophic archaebacteria growing optimally at 100°C. *Arch. Microbiol.* **145:** 56–61.

Lin, C.C. and L.E. Casida. 1984. Gelrite as a gelling agent in media for the growth of thermophilic microorganisms. *Appl. Environ. Microbiol.* **47:** 427–429.

Moorhouse, R., G.T. Colegrove, P.A. Sandford, J.K. Baird, and K. Imahori. 1981. PS-60: A new gel-forming polysaccharide. *ACS Symp. Ser.* **150:** 111–124.

Zillig, W., I. Holz, D. Janekovic, W. Shäfer, and W.D. Reiter. 1983. The archaebacterium *Thermococcus celer* represents a novel genus within the thermophilic branch of the archaebacteria. *Syst. Appl. Microbiol.* **4:** 88–94.

PROTOCOL 4

Continuous Culture Techniques for Thermophilic and Hyperthermophilic Archaea

R.N. Schicho, S.H. Brown, I.I. Blumentals,
T.L. Peeples, G.D. Duffaud, and R.M. Kelly

Described here is a relatively simple system developed for the cultivation of extremely thermophilic and hyperthermophilic anaerobes and aerobes. Although this system was originally used to grow *Pyrococcus furiosus* (Brown and Kelly 1989), it has since been utilized in our laboratory and several others to cultivate a variety of thermophiles, including members of the genera *Pyrodictium, Thermotoga, Thermococcus,* and *Methanococcus,* as well as ES4 and other new isolates. The same system, run under aerobic conditions, has been used to grow the aerobic thermoacidophile *Metallosphaera sedula* (see, e.g., Peeples and Kelly 1994). Applications include the determination of growth parameters, such as yield coefficients (see, e.g., Schicho et al. 1993a), information on metabolic regulation (see, e.g., Snowden et al. 1992; DiRuggiero et al. 1993; Schicho et al. 1993b), and generation of biomass for enzyme research (see, e.g., Schuliger et al. 1993).

Reactor System

A schematic of the experimental system is shown in Figure 1. The culture vessel is a 5-neck round-bottom flask (Lab Glass LG-7390) with a total volume of 2 liters; larger volumes can be used. A gas dispersion tube is used to sparge the vessel. The effluent gas from the headspace exits the vessel through a Graham condenser to reduce water loss. The effluent gas then passes through a manifold from which it may be sampled by a mass spectrometer or gas chromatograph and is then vented to a hood. For cultures grown on elemental sulfur that produce large amounts of sulfidic gases, scrub the effluent gas with an NaOH solution to reduce odor and minimize sulfide sent to the vent. Agitation is provided by an egg-shaped stir bar and stir plate. Maintain the temperature in the culture vessel using a heating mantle (Glas-Col, Cole-Parmer G-36225), proportional temperature controller (Cole-Parmer G-02186), and a type J thermocouple (Cole-Parmer G-08146). The pH of the culture may be monitored and/or controlled using a miniature, autoclavable, double-junction pH electrode. Use a pH controller (Cole-Parmer G-05652) to add 1 N HCl or H_2SO_4 as needed.

Continuous operation is performed with a single, multihead peristaltic pump using tubing of different internal diameters such that the maximal outlet flow rate is always faster than the sum of all inlet flow rates (medium and colloidal sulfur, if added). The outlet tubing is connected to the culture vessel by an overflow tube so that the culture volume is maintained at 1000 ml or whatever working volume is chosen. For example, a Masterflex peristaltic pump (Cole-Parmer 07520-00) may be used with a size-14 pump head for addition of colloidal sulfur, size-16 pump head for addition of culture medium, and size-15 pump head for effluent removal; if all three are attached to a single pump drive, then the medium and colloid added through the size-14 and size-16 tubings will never exceed that removed through the size-15 tubing regardless of dilution rate. Norprene tubing is used for its low O_2 permeability, autoclavability, and durability. Connections are made with polypropylene Luer-Lok fittings (Cole-Parmer G-06359). Glass tubing connections, pH electrode, thermocouple, and gas dispersion tube are sealed to the flask necks with Teflon thermometer adapters with Viton O-rings (Cole-Parmer G-06832).

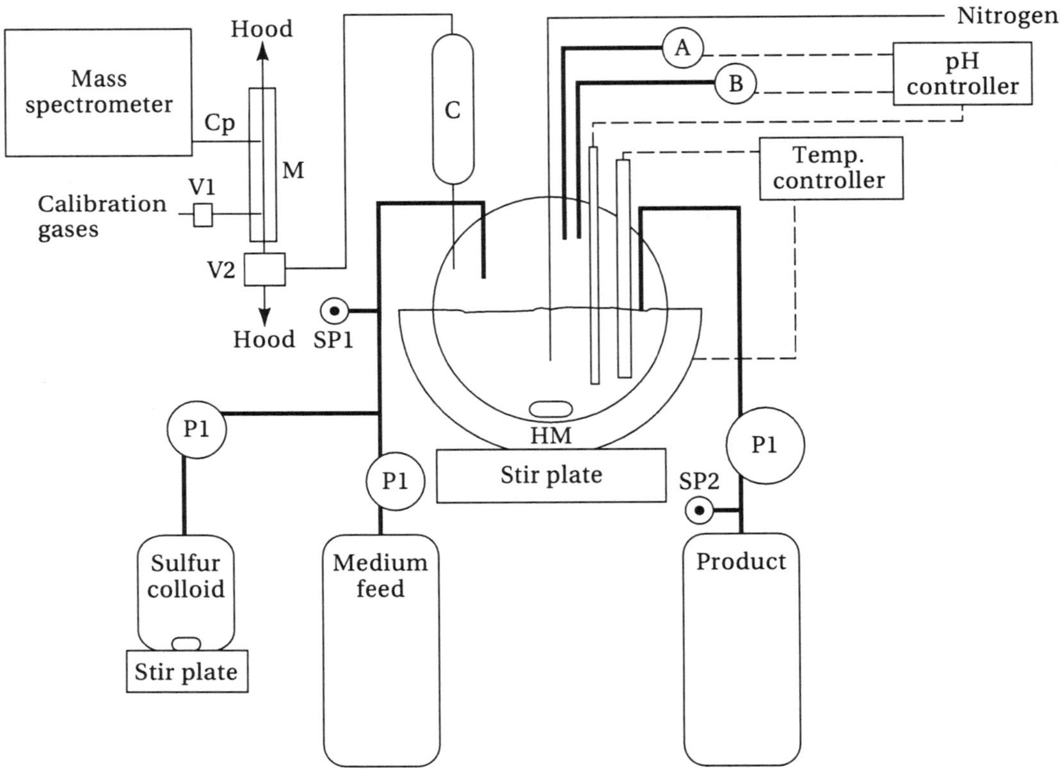

Figure 1 Continuous Culture Apparatus. (*Thin line*) Gas line; (*heavy line*) liquid line; (*dashed line*) signal line. (SP) Sample point; (P1) main pump; (A) acid pump; (B) base pump; (HM) heating mantle; (C) condenser; (V1) gate valve; (V2) three-way valve; (Cp) capillary; (M) manifold.

Preparation of Colloidal Sulfur for Continuous Feed

Preparation of sulfur colloid was from a procedure described by Janek and summarized by Schicho et al. (1993a). The colloid may not be sterilized by autoclaving; however, the feed bottle with enough water to dilute the colloid for use may be autoclaved and cooled prior to adding the concentrated colloid and connecting to the culture system. The colloid feed is sparged with N_2 and stirred during continuous culture experiments.

MATERIALS

H_2O (distilled)
H_2SO_4 (concentrated)
Magnetic stirrer (Cole-Parmer 04768)
Solution A: 64 g of $Na_2S{\cdot}9H_2O$ in 500 ml of distilled H_2O
Solution B: 36 g of Na_2SO_3 in 500 ml of distilled H_2O

Caution: Sulfide is extremely toxic. Wear gloves and safety glasses and work in a chemical fume hood.

METHODS

1. Add 15 ml of Solution B to Solution A, and then add 30 ml of concentrated H_2SO_4 to Solution B.
2. Slowly add 80–100 ml of diluted H_2SO_4 (15 ml of acid + 100 ml of distilled H_2O) to Solution A until turbitity persists.
3. Slowly add (30–160 seconds) Solution A to Solution B with vigorous stirring and then let stand for 1 hour.
4. Dilute the suspension to 2 liters with distilled H_2O and then add 20 g of NaCl to promote settling.
5. Stir the suspension to dissolve the NaCl and then let stand 4 hours to overnight. The colloid will settle to less than 50 ml volume.
6. Remove the slightly cloudy solution by aspiration and replace with 2 liters of NaCl solution (10 g/liter). Resuspend (with a magnetic stirrer) and again allow the suspension to settle.
7. Repeat this washing procedure twice and then resuspend the colloid to 1 liter volume with distilled H_2O and store at room temperature.
8. Dilute the suspension with distilled H_2O (10–20x) to a turbidity of $A_{850} = 1.25$ for use in continuous culture. The suspension will remain useful for several months, but the particles will begin to agglomerate after 1 month and will not fully resuspend.

Cultivation of Thermoanaerobes

Pyrococcus furiosus was grown in artificial sea water (Medium 4), supplemented with yeast extract and tryptone. *P. furiosus* has an obligate requirement for a source of peptides and can be grown in a medium composed of the ASW supplemented only with yeast extract (0.1%) and tryptone (0.5%). It can also be grown in media supplemented with minimal amounts of yeast extract and tryptone (0.01% each) and maltose or a polysaccharide (e.g., starch) as the main carbon source. In such a semidefined medium, optimal growth requires the addition of a trace element and a vitamin solution. Maximal growth requires 0.5% maltose in such a medium. Anaerobic conditions must be maintained throughout, and resazurin is used at a concentration of 1.0 mg/liter of medium as an indicator of anaerobicity.

Anaerobicity is most conveniently obtained by sparging the medium with N_2 and adding cysteine-HCl at a concentration of 2.5 g/liter. Other reducing agents, such as Na_2S or Ti(III) citrate, may be used but frequently result in the formation of precipitates in the medium.

MATERIALS

Growth medium and solutions for *Pyrococcus furiosus* (see Appendix 2, Medium 4)
Artificial seawater (ASW; modified from Kester et al. 1967) supplemented with yeast extract and tryptone (Difco 0127)
Sulfur powder (Baxter 8400-500NY)
Serum bottles (125 ml; Fisher 06-406K)
Butyl rubber stoppers (Fisher 06-406-14A)
Aluminum crimp seals (Fisher 06-467E)
Polypropylene carboys (Nalge 02-960-20)
Filters (0.2 μm) (Gelman, Acrodisk, Fisher 09-730-264A)

METHODS

1. Grow inocula for continuous cultures in 125-ml serum bottles containing 50 ml of medium and 0.5 g of elemental sulfur powder.

2. Prepare vials for inoculation by preheating to 98°C, sparging with N_2, adding cysteine-HCl to 0.25%, and sealing with butyl rubber stoppers secured with aluminum crimp seals.

3. Once clear (indicating anaerobicity), inoculate the medium from a storage culture with approximately 0.5 ml. A bottle culture grown in this manner may be stored at 5°C for several months as a source of inocula. Typically, use 10–50 ml of late log phase culture (~8 hours) from such a vial to inoculate the 1000-ml working volume of the continuous culture flask. The medium in the flask being likewise pre-heated, sparged with N_2, and made anaerobic.

4. Start continuous operation during late log phase. If sulfur powder is being used, add additional sulfur as it is depleted from the reactor, the rate varying with the dilution rate.

5. Autoclave medium in polypropylene carboys for 1 hour. Sparge the medium with N_2, after cooling to 60–80°C, and make anaerobic with cysteine-HCl at this time. Make any other additions (e.g., maltose and vitamin solution) after autoclaving. Filter all additions (gas and liquid) through 0.2-μm filters to maintain sterility.

Cultivation of Thermoacidophiles

Metallosphaera sedula (DSM 5348) cells are cultivated in a basal medium containing the 0.4 g/liter K_2HPO_4, 0.4 g/liter NH_4SO_4, and 0.4 g/liter $MgSO_4 \cdot 6H_2O$. This medium is acidified to pH 2.0 by adding 1 ml of 10 N H_2SO_4 per liter of salts mixture. The continuous culture apparatus is the same as that used for thermoanaerobes except that the steel internal parts are replaced with Teflon-coated parts to avoid corrosion at low pH. In addition, the N_2 sparge gas of the anaerobic fermentation is replaced with an air sparge or, in some cases, with a mixture of air and 10% CO_2 in nitrogen.

MATERIALS

Growth medium and solutions for *Metallosphaera sedula* are identical to those for *Sulfolobus* (see Appendix 2, DSM 88 Medium)
Round-bottom flasks (Lab Glass LG-7371)
Filters (0.2 mm) (Gelman, Acrodisk, Fisher 09-730-264A)

METHODS

1. Autoclave sterile basal media, supplemented with appropriate organic substrates (0.2% yeast extract or casamino acids), add media to the flask, and then supplement with inorganic substrate (10 g/liter $FeSO_4 \cdot 7H_2O$, S^o, or FeS_2).

2. Heat the flask to culture temperature (75°C) and aerate before inoculation.

3. Start continuous culture by pumping fresh medium containing salts and organic substrates (pH 2) and a slurry of inorganic substrate (pH 2) into a round-bottom flask containing cells at mid log phase.

4. To avoid contamination of the culture, filter the inorganic feed through a 0.2-mm filter. The dilution rate for most experiments is 0.05 hr^{-1}. Although pH control is possible with this system, no attempts have been made as yet to control the pH of thermoacidophilic continuous cultures.

COMMENTARY

• The addition of elemental sulfur to the growth medium for *P. furiosus* results in a stimulation of the maximal biomass yield obtained under carbon-limited conditions in continuous culture and facilitates batch growth in serum vials (Schicho et al. 1993). Elemental sulfur may be added in powder form to the reactor vessel or may be delivered on a continuous basis as a colloid. The colloid agglomerates and settles rapidly at NaCl concentrations above 10 g/liter and therefore must be added to a continuous culture separately from the medium.

• The medium described (ASW plus trace elements solution) was prepared and autoclaved at 25% concentrated to allow for the dilution by the delivery of the colloid at 25% of the total volumetric flow rate (medium plus colloid). Under sulfur-limiting conditions, the very low concentrations of sulfur present in the reactor probably exist as polysulfides (Blumentals et al. 1990) due to the presence of the product of sulfur metabolism, H_2S, and no sulfur colloid accumulates in the reactor.

REFERENCES

Blumentals, I.I., M. Itoh, G.J. Olson, and R.M. Kelly. 1990. The role of polysulfides in the reduction of elemental sulfur by the hyperthermophilic archaebacterium, *Pyrococcus furiosus. Appl. Environ. Microbiol.* **56:** 1255–1262.

Brown, S.H. and R.M. Kelly. 1989. Cultivation techniques for hyperthermophilic archaebacteria: Continuous culture of *Pyrococcus furiosus* at temperatures near 100°C. *Appl. Environ. Microbiol.* **55:** 2086–2088.

DiRuggiero, J., S.H. Brown, L.A. Achenbach, R.M. Kelly, and F.T. Robb. 1993. Regulation of ribosomal RNA transcription in the hyperthermophile, *Pyrococcus furiosus*: Evidence for a stringent response. *FEMS Lett.* **111:** 159–164.

Kester, D.R., I.W. Duedall, D.N. Connors, and R.M Pytkowicz. 1967. Preparation of artificial seawater. *Limnol. Oceanogr.* **12:** 176–178.

Peeples, T.L. and R.M. Kelly. 1994. Bioenergetics of the metal/sulfur oxidizing extreme thermoacidophile, *Metallosphaera sedula. Fuel* **72:** 1619–1624.

Schicho, R.N., K. Ma, M.W.W. Adams, and R.M. Kelly. 1993a. Bioenergetics of sulfur reduction in the hyperthermophilic archaeum, *Pyrococcus furiosus. J. Bacteriol.* **175:** 1823–1830.

Schicho, R.N., L.J. Snowden, S. Mukund, J.B. Park, M.W.W. Adams, and R.M. Kelly. 1993b. Influence of tungsten on metabolic patterns in the hyperthermophile *Pyrococcus furiosus. Arch. Microbiol.* **159:** 380–385.

Schuliger, J.W., S.H. Brown, J.A. Baross, and R.M. Kelly. 1993. Purification of a novel amylolytic enzyme from ES4, a hyperthermophilic archaeum. *Mol. Marine Biol. Biotechnol.* **2:** 76–87.

Snowden, L.J., I.I. Blumentals, and R.M. Kelly. 1992. Regulation of intracellular proteolysis in the hyperthermophilic archaebacterium *Pyrococcus furiosus. Appl. Environ. Microbiol.* **58:** 1134–1141.

PROTOCOL 5

High-pressure–High-temperature Cultivation of Thermophilic Archaea

C.M. Nelson and D.S. Clark

The following protocol describes the use of high-pressure reactors for culturing extreme thermophiles and thermophilic enzymes. The systems have been used to study *Methanococcus jannaschii* (Miller et al. 1988a,b) and the hyperthermophilic *Pyrococcus endeavori* (previously identified as ES4) (Nelson et al. 1991, 1992).

Methanococcus jannaschii

Methanococcus jannaschii (DSM 2661) was obtained from the Deutsche Sammlung von Mikroorganismen, Germany. The culture was maintained in artificial seawater (ASW) buffered with PIPES (piperazine-*N,N'*-bis[2-ethanesulfonic acid]) and passaged approximately once every other month.

The high-pressure reactor system has been described previously (Miller et al. 1988a,b; Nelson et al. 1991) and is depicted in Figure 1. The reactor vessel is constructed of 316 stainless steel. Hyperbaric pressurization is employed to provide a continuous supply of substrate gases. In most cases, helium is used as a pressurizing gas. The major components and a step-wise procedure for operation of the reactor follow.

MATERIALS

Growth medium for high-pressure cultivation of *M. jannaschii* (see Appendix 2, Medium 5)
Helium (industrial grade 99.995% minimum purity; Altair UN1046)
EtOH/deionized H_2O solution (70%)
H_2O (sterile, deionized)
Reactor system parts list (see Fig. 1)
- Parts listed below are from High Pressure Equipment Co.
  - Pressure vessels No. 1 (R1.5-6-30) and No. 2 (MS-17)
  - Pressure generator PG (50-6-15)
  - Automated sample valves A1 and A2 (30-11HF4, normally closed) and A3 (30-11HF2, normally open)
  - Check valve CV1 (60-41HF2)
  - High-pressure tubing (1/8-inch outer diameter x 0.040-inch inner diameter and 1/4-inch outer diameter x 0.083-inch inner diameter, 316 stainless steel) and valves (1/8-inch outer diameter [30 HF2] and 1/4-inch outer diameter [30 HF4])

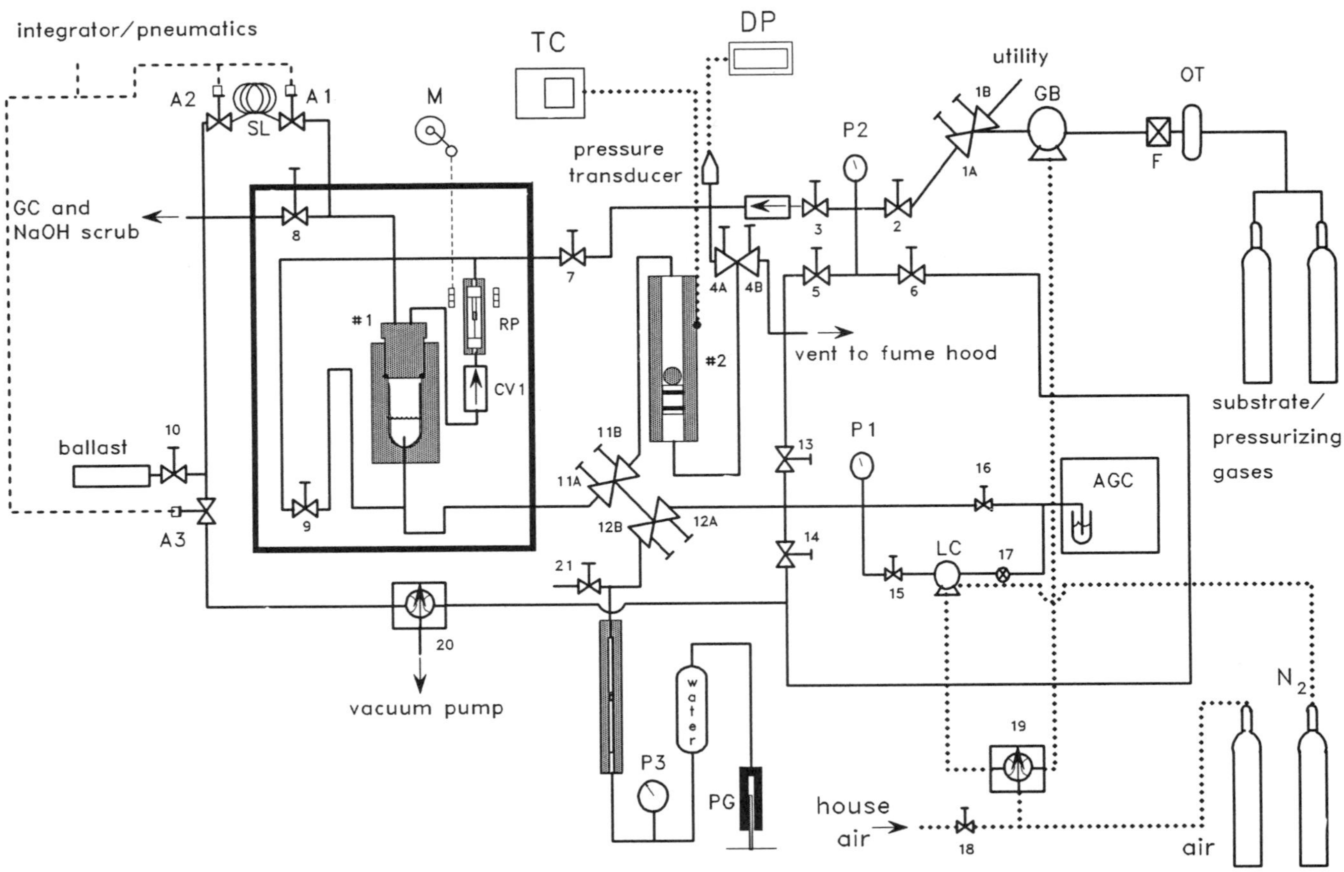

Figure 1 High Pressure System Schematic. (DP) Digital pressure gauge; (EC) electronic temperature controller; (M) motor; (OT) oxygen trap; (GC) gas chromatograph; (AGC) anaerobic glove chamber; (PG) pressure generator; (GB) gas booster; (LC) liquid compressor; (RP) recirculation pump; (F) filter; (SL) gas sample loop.

Gas booster GB (Haskel AG-152)
Liquid compressor LC (Haskel ASF-150)
Anaerobic glove chamber AGC (C100A, Coy Laboratory Products)
Laboratory oven (Blue M OV-4904-2)
Temperature controller TC (Omega Engineering 11-463-44)
High-pressure syringe/Gas recirculation pump. Both were custom-manufactured and constructed from 316 stainless steel to resist corrosion due to native oxide layer (see Commentary).

Caution: High-pressure gas can cause rapid suffocation. Always use equipment rated for maximum possible system pressure.

METHODS

1. Sterilize reactor, inlet lines, and sampling system by pumping a 70% EtOH/deionized H_2O solution from the anaerobic glove chamber (AGC) into the reactor and out through the sampling system.

2. Increase oven temperature to 86°C and evacuate reactor system by opening valves to vacuum pump. Purge system once with helium and re-evacuate.

3. Aseptically add sulfur source and reducing agent to 100 ml of sterile medium. Heat medium to 80°C in temperature-controlled water bath with agitation.

4. Fill reactor system with 50 psig substrate gas. Allow 1 hour for the temperature to equilibrate.

5. Pump 20 ml of uninoculated medium through inlet lines and out to the chemical fume hood. With gas recirculation off, pump 60 ml of medium into reactor (reactor pressure should increase to ~82 psig).

6. Start gas recirculation in hyperbaric reactor and slowly (≤10 psi/sec) add helium until the desired final pressure is reached.

7. Flush sterile, deionized H_2O through liquid inlet lines out to the fume hood to prevent buildup of salts and sulfide in the transfer lines. Allow reactor to remain at cultivation temperature overnight to scavenge trace amounts of O_2 and prereduce metal surfaces.

8. Vent sterile medium from reactor. Restore system pressure to 50 psig with substrate gas. Allow 1 hour for temperature equilibration.

9. Aseptically add sulfur source and reducing agent to 100 ml of sterile medium. Heat medium to 80°C in temperature-controlled water bath with agitation.

10. Aseptically inject 1 ml of stock inoculum (*M. jannaschii*, 10^8 cells/ml) into preheated medium. Carry inoculated medium into anaerobic glove chamber.

11. Pump 20 ml of inoculated medium through inlet lines and out to the fume hood. With gas recirculation off, pump 60 ml of medium into reactor (reactor pressure should increase to ~82 psig). Seal remaining medium in vial and return to water bath. Use remaining medium as a positive control for growth. If no growth is apparent in the reactor, the presence or absence of growth in the bottle will help to locate the problem. Replace headspace in bottle with substrate gas.

12. Start gas recirculation in hyperbaric reactor and slowly (≤10 psi/sec) add helium until the desired final pressure is reached.

13. Flush sterile, deionized H_2O through liquid inlet lines out to fume hood to prevent buildup of salts and sulfide in the transfer lines.

COMMENTARY

- The reducing conditions required for growing thermophilic anaerobes can degrade the native oxide and increase the likelihood that metals will leach into the medium. A stainless steel vessel had no noticeable effect on the growth or metabolism of *M. jannaschii* (Miller et al. 1988a), but it inhibited and altered the metabolism of *P. endeavori* (Nelson et al. 1991). Contact with stainless steel can be minimized by incorporating a glass liner or electroplating the interior surfaces with 24k gold. If cultures are maintained under oxidizing conditions, passivation of the metal surface with nitric acid has been shown to reduce growth inhibition (Sonnleitner et al. 1982).

• Growth and metabolism can be monitored by gas chromatographic (GC) analysis of the reactor headspace and by measuring protein concentration (Bio-Rad microassay technique; Miller et al. 1988b) of periodically withdrawn liquid samples. Production of CH_4 and consumption of substrate are growth-associated under most conditions.

Pyrococcus endeavori

P. endeavori was obtained from Robert M. Kelly (North Carolina State University, Raleigh). *P. endeavori* was maintained in carbonate-buffered artificial seawater (ASW, see Appendix 2, Medium 10). The culture was passaged approximately once a month. The medium used for cultivating *P. endeavori* was described by Brown and Kelly (1989) and is prepared as follows:

METHODS

Individual solutions are prepared separately and sterilized for storage. Prior to use:

1. Mix equal volumes of Solutions A and B. Adjust pH to 6.0 with HCl.

2. Add yeast extract (1 g/liter), tryptone (5 g/liter), and resazurin (0.1 ml/liter of 10 g/liter solution). Purge resulting medium with N_2 for 20 minutes per liter of medium (min/liter).

3. Seal medium under N_2 in anaerobic sample vials (125 ml per vial). Autoclave for sterility.

4. Add substrate (Solution C, 2.5 ml) and reducing agent (5% Na_2S solution, 1.25 ml) to each vial of preheated medium just prior to inoculation.

Hyperbaric Pressurization

The reactor is shown in Figure 1 and is described elsewhere (Miller et al. 1988a,b; Nelson et al. 1991). Hyperbaric pressurization is accomplished as described above for *M. jannaschii* with the following exceptions:

1. Elemental sulfur (S^o) is necessary for good growth and must be provided in the reactor system. Add 1–3 g of S^o (orthorhombic) to sterile glass insert (Fig. 2).

2. Slide insert into presterilized reactor.

3. Optimum growth temperature is 98°C, so increase the reactor temperature accordingly.

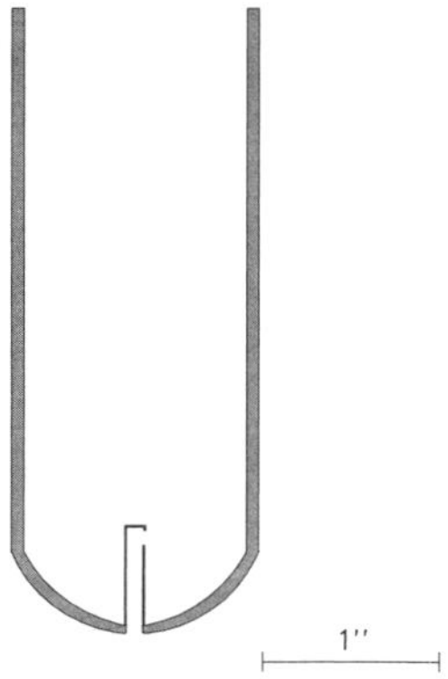

Figure 2 Pyrex Reactor Insert. The tube in the center of the insert is for addition and removal of medium.

4. Use 2 ml of inoculum per 100 ml for *P. endeavori.*

5. No gas-phase substrates are required so only helium is required in the head-space.

6. Avoid agitation/gas recirculation except immediately prior to gas or liquid sampling.

7. Bio-Rad microassay technique is not effective for measuring growth due to the presence of resazurin and initial protein in the preinoculated medium. Monitor growth by counting cells periodically in withdrawn liquid samples.

8. Production of CO_2 and H_2S is indicative of metabolic activity.

Hydrostatic Pressurization

Reactor vessel No. 2 of Figure 1 is used for hydrostatic experiments with a Teflon piston separating the pressurizing gas from the culture (Nelson et al. 1992). The hydrostatic vessel is inoculated in a manner similar to that described above with some minor modifications:

1. Manually position piston and mixing ball near the top of vessel No. 2 allowing room for 1.5 g of S^o.

2. Add 0.5–1.5 g of S^o and reassemble the reactor. Purge and sterilize reactor system as described above for hyperbaric pressurization. Preheat reactor to 98°C (or desired temperature) with electronic temperature controller. Prior to inoculating the reactor, scavenge trace O_2 and prereduce metal surfaces by adding sterile medium with reducing agent (same procedure as for inoculation) and equilibrating overnight at the cultivation temperature.

3. Inoculate and preheat medium as described above for hyperbaric pressurization.

4. Expel sterile medium from reactor through liquid sample line. Maintain pressure behind piston by adding helium.

5. After flushing liquid inlet lines with medium, add 50 ml to the hydrostatic vessel. (Liquid volume added is measured directly, not by relative pressure increase.) Vent excess pressure to fume hood during addition of liquid or equilibrate gas phase with reactor No. 1 to allow adequate room for liquid addition.

6. Expel accumulated gas in liquid side prior to application of final pressure.

7. Slowly add helium ($\leq$10 psi/sec) to gas side until desired pressure is reached.

8. Magnetically lift and release mixing ball prior to liquid sampling.

Gas and Liquid Sampling

Gas samples are removed from the hyperbaric system in one of two ways, manually or automatically. The main considerations during sampling involve the total pressure and sample volume.

Gas Sampling for Pressures $\geq$50 atm

Automatic Sampling

1. Install 0.3 cm^3 sample loop (SL in Fig. 1).

2. Purge gas from the sample line before gas analysis (the lines between no. 1 and valves 8 and A1). The tubing has a small inner diameter so that diffusion dominates mixing and is too slow for the composition in the gas line to be representative of the gas composition in the vessel. The sample loop has a volume approximately equal to the volume of the sample line, so one sample is sufficient to purge the line.

3. The purge step removes enough gas to fill the gas chromatograph (GC) lines several times over. To protect the GC, open valve 10 to the ballast vessel to allow decompression of the gas before entrance into the GC sampling system.

4. Evacuate gas lines and sample loop.

5. Withdraw gas sample and start gas chromatograph analysis.

Manual Sampling

During manual sampling, there is a potential for rapid release of system pressure. NaOH scrubber should be adequately covered to prevent splashing.

1. Slowly open sample valve 8 until vessel pressure starts to drop.

2. Allow gas to flow until pressure has dropped 0.5%. Close valve and evacuate gas lines.

3. Slowly open sample valve again until pressure starts to drop. Allow gas to flow until gas chromatograph lines are filled and gas bubbles are seen to exit through NaOH scrubber.

4. Start gas chromatograph analysis. Valve 10 to ballast tank need not be opened during manual sampling.

Gas Sampling for Pressures <50 atm

The procedures are similar to those above except for the following.

1. Evacuate gas chromatograph lines prior to sampling.

2. Install 2.0-cm^3 sample loop for automated sampling (or 1.5-cm^3 loop requiring two samples) to provide enough gas to fill gas chromatograph lines.

3. No purge is necessary since sample volume is much greater than the volume between valves 8 and 1.

4. Valve 10 to ballast vessel should not be opened. For manual sampling, withdraw enough gas to fill lines. No extra purge is necessary.

Liquid Sampling

When removing liquid samples from the high-pressure system, take care to prevent accidental release of system pressure. If intact cells are required (as with ES4), and the total pressure is ≥50 atm, low-shear sampling and slow decompression are necessary. This is accomplished with a high-pressure syringe (see Nelson et al. 1991) as described below.

1. Prepare a number of 15-ml anaerobic tubes (at least as many as desired liquid samples). Wash the tubes with acid (2% HCl soak), dry, seal in the anaerobic glove chamber, and autoclave for sterility.

2. Open recirculation valve (9) and turn on recirculation pump (if off) for 1 minute.

3. Attach a sterile needle to the sample exit valve (21) for injection into the sample vials. Close the recirculation valve and turn off the recirculation pump motor.

4. Pressurize sampling system to a pressure equal to or slightly greater than the reactor system pressure. Draw approximately 1 ml into the syringe (liquid should be removed from the vessel by backing out the pressure generator) to purge lines.

5. Close liquid sampling valve 11 and reduce pressure in syringe by backing out pressure generator. Although only a purge, drop the pressure relatively

slowly (<100 psig/sec). Once the pressure reads atmospheric (P3), open the sample valve (21) slightly to vent expanded gases. *Note:* Expanded gases may cause the required decompression volume to exceed the syringe volume. Once the piston has traveled the full length of the syringe, the measured pressure will not be representative of the sample pressure, which can still be as high as several hundred psig.

6. Expel liquid into a waste tube, close valve 21, and increase syringe pressure back up to reactor pressure.
7. Draw 1.5 ml into sampling syringe.
8. Slowly decompress sample allowing sufficient time for evolution of dissolved gases (0.5–1-minute pause after each 0.025 increase in volume).
9. Once the piston has traveled the full length of the syringe, final decompression is accomplished by slow release of expanded gases through sample valve.
10. Inject liquid sample into sterile vial.

COMMENTARY

• After sampling, the sample can either be fixed for microscopy, refrigerated for later subculture, or subdivided for both. If the system pressure is <50 atm, or whole cells are not required (as for total protein assays), samples can be removed directly from the reactor without decompression. Sample lines still need to be purged, however.

• Duplicate experiments can be performed in series or a second set of experimental conditions (temperature and pressure) can be investigated without having to clean and service the reactor system. After collecting data for a single run, the system can be emptied of most of the liquid and refilled with fresh, preheated medium. Depending on the particular experiment, the fresh medium can be sterile or inoculated. The temperature and pressure can be reset or remain constant. Although it is theoretically possible to sterilize the system in situ, complete disassembly and sterilization are recommended after 5 days of operation or whenever a different organism is to be cultured. The hyperbaric and hydrostatic reactors can be operated in parallel enabling direct comparison of the two modes of pressurization.

REFERENCES

Balch, W.E. and R.S. Wolfe. 1976. New approach to cultivation of methanogenic bacteria: 2-Mercaptoethane sulfonic acid (HS-COM)-dependent growth of *Methanobacterium rumingutium* in a pressured atmosphere. *Appl. Environ. Microbiol.* **32:** 781.

Brown, S.H. and R.M. Kelly. 1989. Cultivation techniques for hyperthermophilic archaebacteria: Continuous culture of *Pyrococcus furiosus* at temperatures near 100°C. *Appl. Environ. Microbiol.* **55:** 2086–2088.

Jones, W.J., J.A. Leigh, F. Mayer, C.R. Woese, and R.S. Wolfe. 1983. *Methanococcus jannaschii* sp. vo. An extremely thermophilic methanogen from a submarine hydrothermal vent. *Arch. Microbiol.* **136:** 254–261.

Miller, J.F., C.M. Nelson, J.M. Ludlow, N.N. Shah, and D.S. Clark. 1989. High pressure-temperature bioreactor: Assays of thermostable hydrogenase with fiber optics. *Biotechnol. Bioeng.* **34:** 1015–1021.

Miller, J.F., N.N. Shah, C.M. Nelson, J.M. Ludlow, and D.S. Clark. 1988a. Pressure and temperature effects on growth and methane production of the extreme thermophile *Methanococcus jannaschii. Appl. Environ. Microbiol.* **54:** 3039–3042.

Miller, J.F., E.A. Almond, N.N. Shah, J.M. Ludlow, J.A. Zollweg, W.B. Streett, S.H. Zinder, and D.S. Clark. 1988b. High-pressure-temperature bioreactor for studying pressure-temperature relationships in bacterial growth and productivity. *Biotechnol. Bioeng.* **31:** 407–413.

Nelson, C.M., M.R. Schuppenhauer, and D.S. Clark. 1991. Effects of hyperbaric pressure on a deep-sea archaebacterium in stainless steel and glass-lined vessels. *Appl. Environ. Microbiol.* **57:** 3576–3580.

——. 1992. High-pressure, high-temperature bioreactor for comparing the effects of hyperbaric and hydrostatic pressure on bacterial growth. *Appl. Environ. Microbiol.* **58:** 1789–1793.

Sonnleitner, B., S. Cometta, and A. Fiechter. 1982. Equipment and growth inhibition of thermophilic bacteria. *Biotechnol. Bioeng.* **24:** 2597.

Zeikus, J.G. 1977. The biology of methanogenic bacteria. *Bacteriol. Rev.* **41:** 514–541.

PROTOCOL 6

Large-scale Growth of Hyperthermophiles

M.W.W. Adams

Some heterotrophic, sulfur-reducing hyperthermophiles such as *Pyrococcus furiosus* (Fiala and Stetter 1986) and *Thermococcus litoralis* (Neuner et al. 1990) can be grown in the absence of elemental sulfur in large-scale (600 liters) stainless steel fermentors. This enables biomass to be obtained in sufficient quantities for large-scale protein purifications. Described here is a specific protocol for growing *P. furiosus*, which is the most studied hyperthermophile to date from the point of view of metabolism and enzymology (Adams 1993, 1994). The medium is a modification of an earlier version (Bryant and Adams 1989) that was adapted from the original methodology of Fiala and Stetter (1986).

MATERIALS

Growth medium for *Pyrococcus furiosus* (see Appendix 2, Medium 6)
Fermentor (600-liter; W.B. Moore)
Sharples centrifuge equipped with stainless steel coiling coils (AS16 or similar model)
Culture flasks (1 liter; Fisher 10-035F)
Glass carboys (20-liter; Fisher 02-887-5)
Glass culture bottles (100 ml; Fisher 06-406J)
Butyl rubber stoppers (Fisher 06-447H)
Syringes (5 ml, 20 ml, and 60 ml; Fisher 14-826-12, 14-823-28, and 14-823-2D)
Vacuum manifold system (e.g., Sargent Walsh 1402)
Incubator (90°C; e.g., Lunaire, model CE-0632)
Argon
Elemental sulfur (S^o)
Na_2S (5%, w/v)

Cautions: Exercise extreme caution in all manipulations with glass culture vessels maintained at 90°C, particularly with the 20-liter vessels. Wear insulated gloves at all times. Check all vessels for hairline fractures before use.

Sulfide is extremely toxic. Wear gloves and safety glasses and work in a chemical fume hood.

METHODS

1. Prepare autoclaved media (50 ml) in 100-ml serum bottles and seal with butyl rubber stoppers. Make these anaerobic by degassing and flushing at least 15 times over a 1-hour period with high purity argon or N_2 (99.99%). Add a small volume (<1 ml) of an anaerobic solution of Na_2S (5%, w/v) by syringe to maintain reducing conditions (evident when the orange-red color of oxidized resazurin disappears).

2. Add a 10% inoculum by syringe and incubate the bottles at 90°C. Growth is monitored by the increase in turbidity at 600 nm. Stationary phase is approached typically after 16 hours when A_{600} = ~0.6.

3. For large-scale cultures, use two 100-ml cultures to inoculate a 1-liter flask containing 500 ml of autoclaved, degassed, and reduced medium.

4. Use two 1-liter cultures as an inoculum for a 20-liter glass carboy containing 16 liters of the same medium. Incubate two 20-liter cultures for 16 hours at 90°C and use both as an inoculum for the 600-liter fermentor.

5. For the 600-liter fermentor, use the same medium but omit S^o, which necessitates sparging with argon (7.5 liters/minute).

6. Maintain the culture in the 600-liter fermentor at 90°C with stirring (50 rpm). Harvest the cells after approximately 16 hours (A_{600} = ~1.0) with a Sharples centrifuge equipped with stainless steel coiling coils at 200 liters per hour.

7. Immediately freeze harvested cells in liquid N_2 and store in a –80°C freezer. The yield of cells is typically 1000 g (wet weight).

COMMENTARY

- *P. furiosus* (DSM 3638) is routinely grown on a small scale as closed static cultures in a medium containing maltose as the carbon source, elemental sulfur (S^o) as the terminal electron acceptor, with yeast extract and tryptone as the nitrogen and vitamin sources, in a medium based on synthetic seawater (Fiala and Stetter 1986) supplemented with tungsten (Bryant and Adams 1989). Cultures can be stored in the same medium at 4°C and remain viable for at least 1 year, although periodic transfer is recommended. *P. furiosus* has a fermentative type of metabolism and produces organic acids, CO_2, and H_2 as final products. When S^o is also added, the reductant that would otherwise be used to generate H_2 is used to reduce S^o to H_2S. This is convenient for small-scale cultures in glass vessels, but S^o-reducing cultures cannot be grown in a large stainless steel fermentor because of the corrosive nature of H_2S. *T. litoralis* (DSM 5473), which also grows well in the absence of S^o, is grown using the same protocol as that described here for *P. furiosus* except that maltose is omitted and the NaCl concentration is 38 g/liter (Neuner et al. 1990). The cell yield is similar to that obtained with *P. furiosus.*

- Perhaps the key to the successful culture of *P. furiosus* is the maintenance of strictly anaerobic and reducing conditions. Resazurin must be added to all media preparations and must remain reduced after inoculation and during growth. Cultures usually tolerate exposure to trace O_2 contamination (indicated by oxidized resazurin) but only at low temperatures (<50°C) and for brief periods. A second important aspect for large-scale cultures concerns when the cells are harvested. These heterotrophic hyperthermophiles appear to have very short stationary phases and this, in the case of *P. furiosus* in particular, is followed by rapid cell lysis. Thus, large cultures must be carefully monitored and cell harvesting (which takes about 2 hours) should be initiated just prior to the end of logarithmic growth to optimize cell yield.

REFERENCES

Adams, M.W.W. 1993. Enzymes and proteins from organisms that grow near and above 100°C. *Annu. Rev. Microbiol.* **47:** 627–658.

——. 1994. Biochemical diversity among sulfur-dependent hyperthermophilic microorganisms. *FEMS Microbiol. Rev.* **15:** 261–277.

Bryant, F.O. and M.W.W. Adams. 1989. Characterization of hydrogenase from the hyperthermophilic archaebacterium, *Pyrococcus furiosus. J. Biol. Chem.* **264:** 5070–5079.

Fiala, G. and K.O. Stetter. 1986. *Pyrococcus furiosus* sp. nov. represents a novel genus of marine heterotrophic archaebacteria growing optimally at 100°C. *Arch. Microbiol.* **145:** 56–61.

Neuner, A., H. Jannasch, S. Belkin, and K.O. Stetter. 1990. *Thermococcus litoralis* sp. nov.: A new species of extremely thermophilic marine archaebacteria. *Arch. Microbiol.* **153:** 205–207.

PROTOCOL 7

Culture of *Thermoplasma acidophilum*

D. Searcy

Because *Thermoplasma acidophilum* can grow in a combination of heat and acidity that few other organisms can tolerate, it can be cultured easily in large quantities and without resort to special fermentors. In nature, it probably respires upon elemental sulfur (Segerer et al. 1988), but since it is facultatively aerobic, large-scale production is easily done with O_2 as the respiratory substrate. Cell production in plastic garbage cans is described below, from which 200 g of cells can be produced every 2 days.

The best source of a *T. acidophilum* culture is to request it from a colleague. Culture collections have not been reliable sources for this organism.

The procedures below are for the type strain *T. acidophilum* 122 1B2 (Darland et al. 1970). Since the cultures may require a minimum threshold inoculum (Smith et al. 1973; and see below), it is not certain that the strain has ever been cloned. Nevertheless, cultures can be purified by growth under selective conditions, of which two types are available: (1) aerobic growth at 59°C and pH <1.5 and (2) anaerobic growth on elemental sulfur at pH 2.5 with accumulation of H_2S. Each strategy is described below. Aerobically and at 59°C, the cultures have a doubling time of approximately 5 hours (Darland et al. 1970).

Small-scale Aerobic Culture

MATERIALS

Growth medium for *Thermoplasma acidophilum* (see Appendix 2, Medium 7A)
Shaking water bath (covered and adjusted to 59°C)
Erlenmeyer flasks (25 ml)
Metal caps for the above flasks, such as aluminum foil
Phase microscope capable of at least 400x

METHODS

1. To start a typical culture, place 10 ml of medium in a 25-ml flask. The medium should be approximately 1 cm deep. Cap the flasks loosely with aluminum foil, heat to boiling, and then cool to 59°C. Inoculate with 0.5 ml of culture. Incubate with shaking (~1.5 cycles/second) at 59°C.

2. Cultures typically grow to maximum density in 2 or 3 days. One day after the cells have reached maximum density, transfer a 5% inoculum to fresh medium. By the fifth day after inoculation, the cells may be mostly dead, and those that survive are likely to be variant types (see below).

COMMENTARY

• Do not substitute other brands of yeast extract for Difco. Most other brands do not support growth of *T. acidophilum* (Smith et al. 1975; D. Searcy, unpubl.).

• It is important to initiate each culture with an inoculum that is at least 5% the volume of the sterile medium (Smith et al. 1973). For example, inoculate 100 ml of medium with 5 ml of active culture. The reason for this is not known.

• Boiling the medium in the flask before inoculation serves two purposes: (1) it sterilizes the flask and (2) it removes excess dissolved O_2 (see O_2 toxicity, p. 58).

• If the cells are not transferred on schedule, the result is declining cell density after each transfer. The cells evidently evolve in culture. Use a transfer schedule that selects for cell types that grow most densely.

• The identity and purity of the cultures can be checked by phase microscopy. In the healthiest cultures, *T. acidophilum* cells are irregular short filaments approximately 0.5–1 μm in diameter. After about 5 minutes at room temperature, the filaments change into spheres of 1–3-μm diameter. In less healthy cultures, the cells are already spherical. Since the filaments (or spheres) are not uniform in diameter, *T. acidophilum* looks quite different from bacterial contaminants. The most common contaminants are actinomycetes, which are long thin cells that are straight, uniform in diameter, and do not round up when cooled.

• Contaminating organisms, when present, can be eliminated either by lowering the pH or by growing the cells on elemental sulfur. *T. acidophilum* can be cultured down to pH 1.0 (Darland et al. 1970), whereas contaminants that can grow below pH 1.7 are seldom encountered. Selective growth on sulfur is described below.

Anaerobic Culture on Sulfur

MATERIALS

Growth medium for *Thermoplasma acidophilum* (except that the pH is 2.5) (see Appendix 2, Medium 7A)

Incubator adjusted to 59°C. An oven type is acceptable.

Serum bottles (20 ml) with rubber stoppers (Fisher 06-406E), or blood collection tubes (Fisher 02-685C), or any other type of container that can be easily evacuated and then hold a vacuum

Hypodermic needle (~23 gauge x 2 cm long) attached to an aspirator or other vacuum source

Silicone grease

Elemental sulfur (S^0). Use "Flowers of Sulfur" or similar powdered form (Fisher 5594-500). Wet the sulfur with about four parts (v/v) 10 mM H_2SO_4. Sterilize by heating for 30 minutes to 100°C. It is normal for the sulfur to float at first, but eventually it should sink. Store it at room temperature, maintaining sterility.

METHODS

1. Prepare medium as described above, except adjust pH to 2.5.

2. Place approximately 6 ml of wetted sulfur in a 20-ml serum bottle. Add medium until the bottle is three-quarters full. Resterilize at this point by heating for 30 minutes at 100°C.

3. When cooled, inoculate with 0.5 ml of *T. acidophilum* culture either from an aerobic culture or from a suspension of sulfur particles with attached cells.

4. Seal the bottle with a rubber stopper. Insert the hypodermic needle through the stopper (rub a small amount of silicone grease on the stopper and on the needle). Evacuate the bottle for 10 minutes; a visible bubbling occurs during this step as O_2 is removed from the sulfur and the medium.

5. After the 10-minute evacuation, remove the needle. To maintain a good vacuum, first put a drop of water on the stopper, surrounding the needle. Then, after the needle has been pulled out, rub a small amount of silicone grease onto the top of the stopper to ensure an air-tight seal.

6. Incubate for 5–14 days at 59°C. Agitation is not necessary, but gentle mixing, such as by rolling, results in better cell growth.

7. To maintain, add fresh liquid medium every 2 weeks. Open the bottle, pour off the old medium, refill with fresh sterile medium, and evacuate as before. The sulfur and adhering cells remain in the bottle.

8. Transfer to aerobic conditions. To transfer a sulfur culture back to O_2, first shake the culture violently to knock the cells off the sulfur particles. Allow the sulfur to settle for 5 minutes. Then take the inoculum from the supernatant, avoiding the sulfur. Transfer a 5% inoculum into a small-scale aerobic culture as described earlier. If after 3–4 days there is no apparent growth, nevertheless transfer a 50% inoculum to fresh, degassed medium and continue aerobic incubation at 59°C.

COMMENTARY

- *T. acidophilum* cultures respiring upon S^0 in sealed bottles can produce up to 15 mM H_2S (Searcy and Hixon 1991). Particularly under acidic conditions, this concentration of H_2S is toxic to virtually all other organisms. Thus, anaerobic growth on sulfur is strongly selective for *T. acidophilum*. New thermoplasma strains can be obtained from natural environments using this type of culture, but add 2 mM Na_2S to the culture at the beginning to inhibit other organisms.

- If large-scale sulfur cultures are grown, caution should be observed because of H_2S toxicity. To humans, H_2S is about equally as toxic as HCN. Be familiar with the indications of H_2S toxicity and its first aid (Subcommittee on Hydrogen Sulfide 1979; Beauchamp et al. 1984). This especially includes artificial respiration for anyone not breathing.

- Once established on sulfur, *T. acidophilum* cultures can be maintained at lower temperatures (e.g., 37°C) and fed fresh medium every 6 months.

- Examination of the cultures is best done by fluorescence microscopy because the cells may be difficult to distinguish from sulfur particles and may be attached (Searcy and Hixon 1991). On a microscope slide, stain a drop of the culture (including sulfur particles) with a drop of acridine orange (10 μg/ml in 4% formaldehyde). Wait 10 minutes, illuminate with green light, and look for red fluorescence.

- Another indicator of cell growth is the accumulation of H_2S. Sterile bottles do not generate H_2S, but a bottle containing an active culture will have H_2S that can be easily detected by smell. Alternatively, colorimetric assays for H_2S can be used, such as the "methylene blue" assay (Fogo and Popowski 1949; Greenberg et al. 1981).

Large-scale Aerobic Cell Growth

For production of cells in large quantity, such as for biochemical preparations, aerobic cultures are preferred because the cells are not mixed with sulfur particles.

MATERIALS

Growth medium for *Thermoplasma acidophilum* (see Appendix 2, Medium 7B)

32-gallon plastic container for the culture (e.g., of fiber-filled polyethylene, polypropylene, or other inert material). We use Sears Permanex garbage cans, which are approved by the National Sanitation Foundation and have a 6-year guarantee. Remove handles and plug holes with silicone stoppers. Do not use Rubber Maid garbage cans; they leach a substance that is toxic to *T. acidophilum.*

Lid for the plastic container above. A wooden board can be used. Cover the lower surface of the board with a polyethylene sheet and staple it in place around the top edge. Alternatively, make a cover from heat-resistant plastic such as polyvinylchloride. Drill at least 10 holes (~20 mm diameter) through the cover: 4 for aeration, 2 for contact thermometers, 1 for immersion heater, 1 for stirrer, 1 for O_2 electrode, and 1 for vent (extend vent hole upward to form a chimney with a PCR pipe 2.5-cm inner diameter x 15 cm long)

Outer containment vessel for security, insulation, and support of the plastic container above. A steel drum cut down to the appropriate height is ideal. We use empty ethanol and acetone drums obtained from the chemical stockroom. Fill the space between the plastic container and the outer steel drum with sand.

Immersion heater (1000 W) jacketed with either glass or ceramic (not stainless steel) (Fisher 11-463-18C)

Two contact thermometers (Fisher 15-180-5) connected in parallel to control the immersion heater through an electronic switch (relay). (The second thermometer is for safety; if either contact closes, the heater shuts off.)

Stirring motor with stainless steel impeller (Fisher 14-503) to stir culture. Bend the impeller so that it forces the medium downward and stirs up the bottom of the culture vessel.

Four gas dispersion tubes (each with coarse fritted glass cylinders; Fisher 11-138-105B) plus four lengths of glass tubing (6-mm inner diameter, 100 cm long; Fisher 11-362-1L) to aerate culture. Using short segments of silicone tubing (5-mm inner diameter) with Y connectors, attach gas dispersion tubes to 100-cm lengths of glass tubing. The purpose of the extensions is to hold the gas dispersion tubes close to the bottom of the culture vessel; otherwise they will float.

Air pump (Fisher 13-875-222)

N_2 gas, used to purge dissolved O_2 from the medium

Polarographic O_2 Sensor (e.g., Yellow Springs Instrument, monitor model 5300 and probe YSI 5331) to measure and regulate O_2 concentration. This instrument is optional (see Methods). The probe is immersed in the culture to a depth of about 30 cm and connected to the O_2 probe meter. The electrical connection to the O_2 probe is run through a 40-cm length of plastic tubing sealed to the back of the probe to protect wires. The recorder output from the probe meter is fed to an electronic switch (see below) that regulates the air pump. The probe is calibrated to 100% equilibrium with air at 59°C by pulling the heated probe out of the bath, exposing it to air, and while it is still hot adjusting the meter to read 100%.

Electronic switch with a voltage-sensitive input circuit that can use output from the O_2 sensor to regulate the air pump. The discriminator on the electronic switch is set to turn on the aeration pump whenever the meter reads less than 50%.

Tangential Flow Filtration Holder and peristaltic pump (Millipore XX42 P2K 60) plus two filter cassettes (0.2-μm pore size; Millipore GVLP 000 05) for harvesting cultures

Prefilter (to remove large particulates) made of four layers of cheesecloth rinsed with water. Tie filter over the top of an 18-cm funnel.

7% NH_4OH. This is 1 volume of concentrated NH_4OH reagent plus 3 volumes of H_2O (up to 400 ml may be needed)

Antifoam A (Sigma A 5633)

METHODS

Inoculum

1. A 115-liter culture requires at least 6 liters of healthy culture inoculum. This is obtained by building up the size of a small-scale aerobic culture, described earlier. Increase the culture volume in several steps, using at least 5% inoculum at each transfer.

2. Finally, the culture will be too large to shake. Use two glass jars or bottles (4 liters) with 3 liters of medium in each. Heat the jars in a water bath and aerate continuously with a glass tube connected to an aquarium pump (no gas dispersion tubes). Incubation is for 2–4 days at 50°C.

3. Check the purity of the inoculum microscopically. Do not use it more than 1 day after it has reached maximum density.

Culture Medium

1. In advance, use the thermostatted immersion heaters to bring 115 liters of water to 59°C. This can take overnight. Stir the container to prevent thermal stratification.

2. The next day, add the medium ingredients (see Appendix 2) and adjust the pH to 1.6. The pH must be measured on a sample cooled to room temperature in an ice bath. Allow 30 minutes at 59°C to hydrolyze the yeast extract and sucrose. As required, add more acid to maintain pH at 1.6.

3. Purge O_2 from the medium with N_2 using a gas dispersion tube. Bottled N_2 gas can be used. Alternatively, place about 100 ml of liquid N_2 in a 500-ml flask and connect it via tubing to the dispersion tube. The important factor is to remove some of the O_2 from the medium, since excess O_2 is toxic to *T. acidophilum*. Upon initial heating, the medium will be supersaturated with O_2.

4. Inoculate a 30-gallon (115 liters) culture with at least 6 liters of healthy culture. Incubate for 24 hours at 59°C with oxygenation regulated to 50% of equilibrium with air. During the first 24 hours, two gas dispersion tubes are sufficient. Aeration is intermittant, and the culture must be stirred with the stirring motor.

5. After 24 hours, the culture should become cloudy with cells and consume O_2 at a rate that causes continuous aeration. The optical density at 540 nm should be approximately 0.2 (measured, e.g., with a Spectronic 20 spectrophotometer, 1 cm light path). Check microscopically for contamination (see above). Approximately 1% contamination can actually result in a higher yield of *T. acidophilum* and be acceptable for certain types of preparations (see Preparing a Cell Lysate). Discard the culture if it becomes too heavily contaminated and start again at a lower pH. We have not experienced contamination below pH 1.6.

6. If all is well after 24 hours, add an additional 1 g of yeast extract per liter. Dry yeast extract placed on top of the culture eventually will dissolve. Do not adjust the pH. At this point, switch the aeration to continuous operation. Use at least four gas dispersion tubes and aerate it as vigorously as possible while maintaining small bubbles. If the culture foams objectionably, add 1 ml of Antifoam A. Remove the stirring motor.

Cell Harvest

By the second day, the culture should have an optical density of about 0.6 and be ready to harvest. If not, delay 1 more day, but no longer. The harvest procedure below, including cleaning up, typically takes approximately 8 hours.

1. Concentrate the cells using the Millipore tangential-flow filtration cell. The culture is drawn from the vessel through the cheesecloth prefilter and then through the peristaltic pump and into the filtration cell. From the exits of the tangential-flow filtration cell, the concentrated "retentate" is returned to the culture vessel and the clear "filtrate" is discarded down the drain. In this way, the volume of the culture is progressively reduced to approximately 20 liters. During this phase of the harvest, aeration must be maintained continuously. Allow the temperature to drop to near room temperature during this filtration step. The culture can be cooled more rapidly by floating in it a stainless steel bucket filled with ice.

2. Partial neutralization. After the culture has been reduced to approximately 20 liters and the temperature is near 25°C, adjust pH to 4.0. Place a pH electrode in the culture vessel. Arrange to mix the culture thoroughly, for example, by violent aeration without using a dispersion tube. The goal is to obtain rapid, top-to-bottom mixing. Slowly add 7% NH_4OH until pH 4 is reached. A good method is to siphon it using 3-mm inner diameter tubing. It can be added as a stream at first, but then reduce the rate to dropwise as pH 4.0 is approached.

   Note: Do not overshoot pH 4.0 and avoid local high concentrations of base! The cells become fragile at higher pH and lyse completely above pH 6 (see Preparing a Cell Lysate). This is not easy because above pH 3, the medium has very little buffering capacity.

3. As the culture approaches pH 4, it will start to foam. Shut off the aeration. Continue the filtration. Transfer the culture to a 10-liter bucket as soon as it will fit, which can be done by diverting the retentate stream into the bucket. After it has been entirely transferred, remove the funnel and cheesecloth prefilter from the intake tubing and replace with a short length of glass tubing. Continue to reduce the volume of the culture by recycling it through the filtration cell until it has been reduced to approximately 1.5 liters.

4. As the volume in the bucket approaches 1.5 liters, save 2 liters of clear filtrate. When the final volume is reached, rearrange the tubing so as to backflush the filters and collect the cells from the inlet side of the filters. In doing this, recover as many cells as possible while not increasing the volume of the cell suspension unnecessarily.

5. Collect the cells by centrifugation at 8000 rpm for 5 minutes at 4°C.

6. Scrape the cell pellets onto a plastic sheet in cookie-like portions and freeze at –70°C.

7. To clean up, flush the filtration cassette with its stack of filters with 1% NH_4OH and then with H_2O. If stronger measures are required, flush with 0.5% sodium dodecyl sulfate (SDS) followed by 6 M urea. Store the filters wetted with 25% glycerol and 1 mM NaN_3 at pH 7. A set of filters should last several years.

   Note: Spilled medium should be sponged up, flooded with water, and then sponged up again. Sulfuric acid is not volatile, and any residue of culture medium will become concentrated H_2SO_4 as the water evaporates. Spatters of nonneutralized medium make holes in clothing. Skin is not badly affected, if reasonable precautions are taken.

Alternative Harvest Procedure

It is possible to harvest the cells using a continuous-flow centrifuge instead of the tangential flow filtration cell. Continuous flow centrifugation cannot be used with hot medium because as it enters the centrifuge head, it immediately convects to the outlets and exits again, without efficient collection of the cells. It is essential to pre-cool the culture. In general, we avoid centrifugal harvests because the medium (containing ammonium sulfate) causes corrosion to the aluminum rotors. We have also had some frightening accidents when the flow distributor at the top the rotor seized up, apparently caused by spilled medium washing away the lubrication.

1. Cool the culture to room temperature by floating it in a bucket filled with ice. Maintain aeration until the culture is fully cooled.

2. Adjust the pH to 4 as described above and proceed with the centrifugation, following the manufacturer's instructions.

COMMENTARY

• A typical yield of wet cells is 1 g/liter, but this may vary from 0.5 to 2 g/liter. When inoculum is saved from one large batch culture to start the next, the yield progressively increases during sequential transfers. "Good" cell pellets are slimy in texture. If the pellets are dry and chalk-like, then the cells have died and are poor material for extracting proteins or nucleic acids.

• *T. acidophilum* is sensitive to lysis either by anaerobiosis, by pH >6, or by traces of detergents. Care must be taken to avoid any of these circumstances. For example, if the filtration cell has been washed with SDS, a residue of detergent remains that causes *T. acidophilum* to lyse within the apparatus. Washing the cell with water is not sufficient; the adsorbed detergent must be flushed out with 6 M urea.

• The organism depends on continuous respiration. Thus, when aerobic cultures become anoxic, the organisms die within 20 minutes (Searcy and Whatley 1984). Dead cells do not exclude acidity, and the nucleic acids and enzymes are quickly damaged.

• Overaeration is harmful, decreasing the yield. Optimal yields are obtained when the O_2 concentration is approximately 50% of equilibrium with air. Nevertheless, it is possible to grow *T. acidophilum* with unregulated aeration. In this case, use larger inocula. Purge excess O_2 from the freshly heated medium, as described earlier. Use about 5 ml air/min/(L medium) for the first day, and thereafter increase it to the maximum rate.

• The gas dispersion tubes should be of fused glass and not aeration "stones" such as found in pet stores. Pet store stones disintegrate in hot acid. In addition, aquarium air pumps from pet stores generally are not powerful enough to aerate a large culture.

• During the first day, the culture must be mechanically mixed with the stirring motor, both to prevent thermal stratification and to ensure that the O_2 probe operates properly. Without mixing, the cooler medium at the bottom is not mixed and becomes anoxic. After the first day, when the culture begins to respire rapidly and is aerated continuously, the bubbling alone provides sufficient mixing.

- For large cultures, sucrose has been substituted for glucose in order to save money. This results in a somewhat lower yield but is cost-effective. Similarly, KOH and H_3PO_4 have been substituted for KH_2PO_4.

- It is a good idea with each batch of Difco yeast extract to test a series of concentrations to find which gives the highest cell density. Too much yeast extract is toxic and lowers the yield of cells.

- Other thermophilic Archaea that are facultative aerobes include the Sulfolobales (Segerer et al. 1986). For example, we have used the above system to grow *Sulfolobus acidocaldarius* (Green et al. 1983). Cross-contamination between the two organisms has not been a problem; apparently, the high concentration of phosphate in *Thermoplasma* medium is toxic to *Sulfolobus*, and the higher temperature used with *S. acidocaldarius* is lethal to *T. acidophilum*.

Storage of Cultures

Although it is possible to maintain *T. acidophilum* cultures on S^o, it is a good idea to have frozen cultures in reserve. For example, there are indications that *T. acidophilum* evolves in culture, and it is unknown what the result of long-term maintenance at low temperature on S^o may be.

MATERIALS

Freezer (–70°C or below, or suitable liquid N_2 container)
NH_4OH (7%)
Sucrose (dry, sterilized by autoclaving)
Sterile containers (2-ml) for freezing (without a constricted neck, such as plastic microcentrifuge tubes; Sigma T 2795)
Ehlenmeyer flask (25 ml)

METHODS

1. Use an aerobic culture 1 day after inoculation. The cultures must have a low cell density (0.05 OD_{540nm}). If necessary, dilute the culture with fresh medium.

2. Cool the culture to room temperature and adjust it to pH 3.0 by adding 7% NH_4OH.

3. Add 0.15 g of sucrose for each 1 ml of culture. Swirl to dissolve.

4. Divide the culture into 1-ml aliquots and freeze at –70°C. It is not desirable to freeze the aliquots either particularly quickly or slowly. Simply place them in a rack in the freezer.

5. Once frozen and stored at less than –70°C, the cultures remain viable indefinitely (apparently).

6. To revive a frozen culture, quickly warm the cells up to 59°C. Preheat 10 ml of medium in a 25-ml flask as described in Small-scale Aerobic Culture (p. 51). Obtain a frozen culture aliquot from the freezer. Immerse it in hot water until the edges melt free of the freezing container, and then drop the culture start, still mostly a pellet of ice, into the preheated medium.

Preparing a Cell Lysate

T. acidophilum has the unique feature of lysis at neutral pH (Searcy 1976). This allows one to open the cells exceptionally gently, without resort to enzymes or detergents. It also means that it is possible to selectively extract *T. acidophilum* lysates from mixed cell cultures. Typically, a 5% lysate is prepared in 10 mM Tris-HCl buffer. This is a good starting material from which to extract proteins and nucleic acids.

MATERIALS

10 mM Tris-HCl buffer (pH 8). This is 2-amino-2-hydroxymethyl-1,3-propanediol adjusted to pH 8 with HCl.
Tris base (1 M; Sigma T 1503)
Glass-Teflon homogenizer (at least 10-ml capacity) (Fisher 08-414-14B)
pH electrode and pH meter. A combination-type single electrode is best.
Centrifuge, refrigerated and capable of 14,000 rpm (20,000*g*)
Sonicator (*optional*)
DNase I (Sigma D 5025) (*optional*)

METHODS

1. Break off and weigh 0.5 g from a frozen pellet of *T. acidophilum* cells. To this add 9.5 ml of 10 mM Tris buffer (pH 8).

2. Transfer to a glass-Teflon homogenizer. By working the pestle up and down by hand as the pellet melts, disperse the cells throughout the buffer. Some lysis will occur at this point.

3. Measure the pH. Add a single drop of 1 M Tris base, mix in the homogenizer, and measure the pH again. Repeat until a pH of 7 is obtained. The lysate should be highly viscous.

4. This step is optional. Sonicate, or otherwise shear, or digest with DNase I until the viscosity is reduced to manageable levels.

5. Centrifuge at 14,000 rpm (20,000*g*) for 15 minutes at 4°C.

COMMENTARY

- The lysate, before sonication, is viscous due to released DNA (Searcy 1976). If it is not viscous, then the cells died prior to being harvested, and the DNA has been destroyed.

- Sonication or digestion results in the DNA being broken into small fragments that remain in solution during the centrifugation step. When it is not sonicated, much of the DNA can be pelleted by high-speed centrifugation (e.g., 18,000 rpm for 30 minutes).

- To prepare *T. acidophilum* membranes and associated respiratory enzymes, avoid neutral pH. Instead, at pH 5.5, rupture the cells using sonication and collect the membrane fraction by centrifugation (see Searcy and Whatley 1982).

REFERENCES

Beauchamp, R.O., Jr., J.S. Bus, J.A. Popp, C.J. Boreiko, and D.A. Andjelkovich. 1984. A critical review of the literature on hydrogen sulfide toxicity. *CRC Crit. Rev. Toxicol.* **13:** 25–97.

Darland, G., T. Brock, W. Samsonoff, and S. Conti. 1970. A thermophilic, acidophilic mycoplasma isolated from a coal refuse pile. *Science* **170:** 1416–1418.

Fogo, J.K. and M. Popowski. 1949. Spectrophotometric determination of hydrogen sulfide. *Anal. Chem.* **21:** 732–734.

Green, G.R., D.G. Searcy, and R.J. DeLange. 1983. Histone-like protein in the archaebacterium *Sulfolobus acidocaldarius. Biochim. Biophys. Acta* **741:** 251–257.

Greenberg, A.E., J.J. Connors, and D. Jenkins, eds. 1981. *Standard methods for the examination of water and wastewater,* 15th ed, pp. 442–448. American Public Health Association, Washington, D.C.

Searcy, D. 1976. *Thermoplasma acidophilum:* Intracellular pH and potassium concentration. *Biochim. Biophys. Acta* **451:** 278–286.

Searcy, D. and W. Hixon. 1991. Cytoskeletal origins in sulfur-metabolizing archaebacteria. *BioSystems* **25:** 1–11.

Searcy, D. and F. Whatley. 1982. *Thermoplasma acidophilum* cell membrane: Cytochrome *b* and sulfate-stimulated ATPase. *Zbl. Bakt. Hyg. I. Abt. Orig. C* **3:** 245–257.

———. 1984. *Thermoplasma acidophilum*: Glucose degradative pathways and respiratory activities. *Syst. Appl. Microbiol.* **5:** 30–40.

Segerer, A., T. Langworthy, and K. Stetter. 1988. *Thermoplasma acidophilum* and *Thermoplasma volcanium* sp. nov. from solfatara fields. *Syst. Appl. Microbiol.* **10:** 161–171.

Segerer, A., A. Neuner, J.K. Kristjansson, and K.O. Stetter. 1986. *Acidianus infernus* gen. nov., sp. nov., and *Acidianus brierleyi* comb. nov.: Facultatively aerobic, extremely acidophilic thermophilic sulfur-metabolizing archaebacteria. *Int. J. Syst. Bacteriol.* **36:** 559–564.

Smith, P.F., T.A. Langworthy, and M.R. Smith. 1975. Polypeptide nature of growth requirement in yeast extract for *Thermoplasma acidophilum. J. Bacteriol.* **124:** 884–892.

Smith, P.F., T.A. Langworthy, W.R. Mayberry, and A.E. Hougland. 1973. Characterization of the membranes of *Thermoplasma acidophilum. J. Bacteriol.* **116:** 1019–1028.

Subcommittee On Hydrogen Sulfide. 1979. *Hydrogen sulfide.* University Park Press, Baltimore.

PROTOCOL 8

Solubilization from Membranes and Purification of Hydrogenases from *Pyrodictium* Species

R.J. Maier, T. Pihl, and J.B. Bingham

The extremely thermophilic *Pyrodictium* species contain membrane-bound H_2-uptake hydrogenases. For the autotrophic bacterium *P. brockii*, this enzyme permits growth with H_2 as the only energy source (Pihl et al. 1990), whereas for *P. abyssi*, H_2 oxidation via hydrogenase facilitates heterotrophic growth (Pley et al. 1991). The nature of these H_2-oxidizing enzymes is consistent with a role for them in energy-conserving electron transport reactions, especially in conjunction with other respiratory membrane-bound electron transport factors (i.e., cytochromes, quinones) (Pihl et al. 1992). The *P. brockii* enzyme has been purified to homogeneity, and the *P. abyssi* enzyme has been partially purified. Even though the molecular (subunit) compositions of the hydrogenase enzymes from the two bacteria are similar, we have found that successful methods for isolation of H_2-oxidizing membranes and detergent solubilization for enzyme purification are markedly different for the two species.

Membrane Preparation and Solubilization of Hydrogenases

MATERIALS

French press and pressure cell (SLM Aminco FA-073)
No. 1 Whatman filter (12.5 cm diameter)
Ultracentrifuge (Beckman L8-70)
Ground glass homogenizer (ice-cold)
Dounce homogenizer (ice-cold)
Rocking Platform (Marsh Biomed Products)
Polypropylene screw-cap tube (50 ml, Corning 25330-50)
Reactive-red 120-agarose (type 3000CL, Sigma R 0503)
Methylene blue (Sigma MB-1)
Wash buffer 1: 50 mM MOPS (3-[*N*-morpholino]propane sulfonic acid) containing 100 mM NaCl, 2 mM EDTA (pH 7.0)
Wash buffer 2: 50 mM MOPS containing 400 mM NaCl, 2 mM EDTA (pH 7.0)
Wash buffer 3: Tween 80 (0.1% w/v; Sigma P 8074) in Wash buffer 1
Detergent buffer 1: Triton X-100 (4% w/v) in Wash buffer 1

Detergent buffer 2: Triton X-100 (2% w/v) in Wash buffer 1
Detergent buffer 3: CHAPS (3-[(cholamidopropyl)-dimethylammonio]-1-propanesulfonate; Sigma C 3023) (6% w/v) in Wash buffer 1
Elution buffer: 50 mM MOPS, 1 M NaCl, 2 mM EDTA (pH 7.0)

METHODS

Pyrodictium brockii

1. Remove large sulfur particles by passing the culture (grown in DSM 283 Medium, see Appendix 2) through a 12.5-cm-diameter No. 1 Whatman filter.
2. Centrifuge cells at 10,000*g* for 20 minutes at 4°C.
3. Wash pellet with Wash buffer 1. Recentrifuge at 10,000 for 20 minutes at 4°C.
4. Homogenize cells in an ice-cold ground glass homogenizer.
5. Pass cells through a French press and pressure cell (two passages at 1250 lb/in^2).
6. Pellet membranes by centrifugation at 110,000*g* for 1 hour at 4°C.
7. Suspend membranes (from a total cell culture of 15 liters) in approximately 5 ml of Wash buffer 1; homogenize with a ground glass homogenizer.
8. To the suspended membranes, add an equal volume of Detergent buffer 1.
9. Incubate on rocking platform (gentle rocking) for 90 minutes at room temperature.

Pyrodictium abyssi

1. Cool freshly grown cells (3–4 liters) to 4°C.
2. Agitate media to resuspend cells plus sulfur particles; transfer to a 4-liter graduated cylinder. Allow sulfur particles to settle.
3. Decant supernatant (cells) and save on ice.
4. Wash sulfur particles in 1 liter of Wash buffer 3 and allow sulfur particles to settle again. Remove supernatant; pool with supernatant from step 3.
5. Centrifuge cells at 10,000*g* for 20 minutes at 4°C and resuspend pellet in Wash buffer 2 (40 ml).
6. Pass cells through a French press and pressure cell (three times at 1250 lb/in^2).
7. Remove unbroken cells and remaining sulfur particles by centrifugation at 5000*g* for 10 minutes.

8. Pellet supernatant (membranes) at 100,000*g* for 1 hour at 4°C.

9. Suspend membranes in 40 ml of Wash buffer 1 and homogenize with an ice-cold Dounce homogenizer; repeat steps 7 and 8.

10. Resuspend final membrane pellet in 10 ml of Wash buffer 1 with ground glass homogenizer.

11. In a polypropylene screw-cap tube, add equal volumes of Detergent buffer 3 and membrane suspension; mix gently.

12. Place tube on rocking platform and allow to rock slowly for 2 hours at room temperature.

13. Pellet membranes at 100,000*g* for 1 hour; remove supernatant containing H_2ase activity. Assay for methylene-blue-dependent H_2 uptake activity (Pihl et al. 1989).

Further Purification of Hydrogenase (*P. brockii*)

This purification procedure is from Pihl and Maier (1991).

1. From step 9 above for *P. brockii*, remove unsolubilized material by centrifugation at 110,000*g* for 1 hour at 4°C.

2. Load solubilized hydrogenase extract (2 ml) onto reactive red agarose (1 x 4-cm column).

3. Wash column with 30 ml of Detergent buffer 2.

4. Wash column with 30 ml of Wash buffer 2.

5. Elute hydrogenase with 15 ml of Elution buffer.

6. Assay fractions for methylene-blue-dependent H_2 uptake activity (Stults et al. 1986; Pihl et al. 1989).

COMMENTARY

Successful solubilization of hydrogenase from *P. abyssi* required a treatment much different from that for the same enzyme from *P. brockii*. From testing a series of detergents, CHAPS (at 3% w/v) was the best detergent for extraction of the *P. abyssi* hydrogenase, with consistent and excellent extraction yields; approximately 100% recovery was obtained compared to the total (washed membranes) activity. Once purified and/or solubilized, the enzyme can be stably stored (–70°C) for long periods or incubated at 0–4°C for short periods.

REFERENCES

Pihl, T.D. and R.J. Maier. 1991. Purification and characterization of the hydrogen uptake hydrogenase from the hyperthermophilic archaebacterium *Pyrodictium brockii*. *J. Bacteriol.* **173:** 1839–1844.

Pihl, T.D., L.K. Black, B.A. Schulman, and R.J. Maier. 1992. Hydrogen-oxidizing electron transport components in the hyperthermophilic archaebacterium *Pyrodictium brockii*. *J. Bacteriol.* **174:** 137–143.

Pihl, T.D., R.N. Schicho, R.M. Kelly, and R.J. Maier. 1989. Characterization of hydrogen-uptake activity in the hyperthermophile *Pyrodictium brockii*. *Proc. Natl. Acad. Sci.* **86:** 138–141.

Pihl, T.D., R.N. Schicho, L.K. Black, B. Schulman, R.J. Maier, and R.M. Kelly. 1990. Hydrogen-sulfur autotrophy in the hyperthermophilic archaebacterium, *Pyrodictium brockii*. *Biotechnol. Gen. Eng. Rev.* **8:** 345–377.

Pley, U., J. Schipka, A. Gambacorta, H.W. Jannasch, H. Fricke, R. Rachel, and K.O. Stetter. 1991. *Pyrodictium abyssi* sp. nov. represents a novel heterotrophic marine archael hyperthermophile growing at 100°C. *Syst. Appl. Microbiol.* **14:** 245–253.

Stults, L.W., F. Moshiri, and R.J. Maier. 1986. Aerobic purification of hydrogenase from *Rhizobium japonicum* by affinity chromatography. *J. Bacteriol.* **166:** 795–800.

PROTOCOL 9

Purification of Hydrogenase and Ferredoxin from a Hyperthermophile

M.W.W. Adams and Z.H. Zhou

Hydrogenase catalyzes the reversible activation of molecular hydrogen (H_2) and has been purified from a range of mesophilic and moderately thermophilic organisms (Adams 1990). The hyperthermophilic archaeon, *Pyrococcus furiosus* (Fiala and Stetter 1986), which grows optimally at 100°C by fermentation and produces H_2, utilizes a cytoplasmic, nickel-containing hydrogenase (Bryant and Adams 1989). A small redox protein termed ferredoxin functions as the physiological electron donor to the hydrogenase (Aono et al. 1989). Both proteins are extremely stable and have several unique properties when compared to their mesophilic relatives (Adams 1992, 1993). The purification procedure to obtain the hydrogenase and ferredoxin of *P. furiosus* has been designed to yield both proteins from the same batch of *P. furiosus* cells, as well as several other enzymes and proteins including aldehyde ferredoxin oxidoreductase, pyruvate ferredoxin oxidoreductase, glutamate dehydrogenase, and rubredoxin (Adams 1993, 1994). The hydrogenase is routinely assayed by H_2 evolution using a reduced redox dye (methyl viologen reduced by sodium dithionite) as the electron donor. The ferredoxin is brown in color and is easily detected by its visible absorption.

MATERIALS

Frozen cells of *Pyrococcus furiosus* (~500 g) grown as described in Protocol 6
FPLC (fast protein liquid chromatography) system (Pharmacia LKB)
Gas chromatograph (model 3300, Varian Associates)
UV-VIS spectrometer (model DMS 200, Varian Associates)
Ultracentrifuge (L8-70, Beckman) with a Ti19 rotor (Beckman)
Shaking water bath (at 80°C)
Vacuum manifold system (e.g., Sargent Walsh, model 1402)
Q-Sepharose column (Pharmacia LKB 17-1014-04)
DEAE-Sepharose Fast Flow column (Pharmacia LKB 17-6708-05)
Mono Q HR 5/5 column (Pharmacia LKB 17-0546-01)
Superose-6 columns (Pharmacia LKB 17-0537-01)
Sepharose CL-6B column (Pharmacia LKB 17-0160-01)
Hydroxyapatite columns (high resolution, Calbiochem-Behring 391948)
Ultrafiltration system (PM-10 Amicon 13132)
Anaerobic stoppered vials (17 x 60 mm, Fisher 03-338C)
Syringes and needles (50 μl, 250 μl, and 500 μl: Fisher 14-815-61, 14-815-97, and 14-815-113)

Cuvettes (1 cm, Uvonic 21NG)
Serum bottles (100 ml, Fisher 06-406J) with stoppers (Fisher 06-406-10B)
DNase I (Sigma DN-25)
Argon
Tris-HCl buffer (50 mM, pH 8.0)
EPPS (*N*-[2-hydroxyethyl]piperazine-*N'*-3-propanesulfonic acid) buffer (50 mM, pH 8.0)
Sodium chloride
Sodium dithionite
Methyl viologen

Caution: Methyl viologen may be fatal if inhaled, swallowed, or absorbed through skin. It is irritating to mucous membranes and upper respiratory tract. Wear gloves and safety glasses and work in a chemical fume hood.

METHODS

Hydrogenase Assay

Hydrogenase activity is routinely determined by H_2 evolution from reduced methyl viologen (1 mM; Bryant and Adams 1989).

1. Degas stoppered vials (8 ml) containing 1 mM methyl viologen and 20 mM sodium dithionite in 50 mM EPPS (pH 8.4) anaerobically and fill with argon.
2. Incubate the vials in a shaking water bath (100 rpm) at 80°C and start the reaction by injecting the hydrogenase sample (10–100 μl).
3. At 1-minute intervals, remove 50 μl of the gas phase and inject into a gas chromatograph to determine the amount of H_2 produced.
4. Calibrate the gas chromatograph using vials maintained at the assay temperature and containing known amounts of H_2.
5. Calculate the hydrogenase activity where one unit catalyzes the production of 1 μmole of H_2 per minute at 80°C. Express the results as units/mg of protein using the Lowry method (Lowry et al. 1951) to estimate the protein content.

Ferredoxin Assay

The brown-colored ferredoxin is followed throughout the purification procedure by its visible absorption at 390 nm, and purity is estimated by the A_{390}/A_{280} ratio.

1. Add an aliquot of the ferredoxin sample (typically 20–100 μl) to a cuvette containing 50 mM Tris-HCl buffer (pH 8.0) in a final volume of 3 ml (performed aerobically).
2. From the absorbance at 390 nm and 280 nm, calculate the A_{390}/A_{280} ratio.
3. Calculate the concentration of the ferredoxin using a molar extinction coefficient at 390 nm of 15,400 M^{-1} cm^{-1}.

Separation of Hydrogenase and Ferredoxin

Hydrogenase activity starts to elute when 0.33 M NaCl is applied to the column, whereas 0.41 M NaCl is required to elute the ferredoxin. The two proteins are then purified separately. All buffers are repeatedly degassed and flushed with argon to remove traces of O_2 and are maintained under a positive pressure of argon.

1. Add 2 mM sodium dithionite to all buffers to protect against trace O_2 contamination.

2. Thaw frozen cells (500 g) under argon with repeated degassing and flushing and then suspend them in 50 mM Tris-HCl buffer (pH 8.0), containing lysozyme (1 mg/ml) and DNase I (0.1 mg/ml) to a total volume of 1.5 liters. Incubate the cell suspension for 1 hour at 30°C with stirring to obtain cell lysis.

3. Centrifuge the suspension using a Ti19 rotor (6 x 250 ml) at 50,000*g* for 40 minutes in an L8-70 ultracentrifuge to obtain the cell-free extract.

4. Apply the cell-free extract directly to a column (5 x 45 cm) of Q-Sepharose previously equilibrated with 50 mM Tris-HCl buffer (pH 7.8) at 8 ml per minute. This and all subsequent columns described below are controlled using an FPLC system.

5. Wash the column with 1.5 liters of 50 mM Tris-HCl buffer (pH 7.8) and elute the absorbed proteins in 90-ml fractions with a 4.2-liter linear gradient from 0 to 0.6 M NaCl at 6 ml per minute in the same buffer.

6. Collect the fractions manually in argon-flushed 100-ml serum-stoppered bottles that have been previously degassed with argon.

Hydrogenase Purification

1. Apply fractions containing hydrogenase activity from the Q-Sepharose column with a specific activity in the H_2 evolution assay (see below) greater than 4 units/mg to a column (2.5 x 30 cm) of DEAE-Sepharose Fast Flow, preequilibrated with 50 mM Tris-HCl buffer (pH 8.3) at 3 ml per minute. The FPLC system is used to dilute the fractions with an equivalent volume of the above Tris-HCl buffer as they are applied.

2. Wash the column with the same buffer (300 ml) and elute the protein in 40-ml fractions with sequential linear gradients from 0 to 0.1 M NaCl (50 ml) and 0.1 to 0.4 M NaCl (1.25 liters) at 2.5 ml per minute. Hydrogenase activity starts to elute as 0.26 M NaCl is applied.

3. Combine fractions with a specific activity above 60 units/mg and apply directly to a column (1.5 x 35 cm) of hydroxyapatite previously equilibrated with 50 mM Tris-HCl buffer (pH 8.0) at 1.5 ml per minute.

4. Wash the column with 90 ml of the same buffer and elute the protein in 5-ml fractions with a 600-ml linear gradient from 0 to 0.20 M potassium phos-

phate in the same buffer (pH 8.0) at 0.8 ml per minute. The hydrogenase elutes as 0.08 M phosphate is applied.

5. Concentrate fractions with specific activity above 120 units/mg to 1.0 ml using a Mono Q HR 5/5 column. Apply the samples to the Mono Q column preequilibrated with 50 mM Tris-HCl buffer (pH 8.0) at 1.0 ml per minute, using the FPLC system to effect dilution with an equivalent volume of the same buffer. Elute the protein at 0.3 ml per minute in 0.5-ml fractions with a linear gradient (3 ml) from 0 to 1.0 M NaCl in the same buffer.

6. Apply the concentrated hydrogenase sample to a column (HR 16/50) of Superose 6 preequilibrated with 50 mM Tris-HCl buffer (pH 8.0) containing 0.2 M NaCl. Run the column at 0.8 ml per minute and collect 5-ml samples.

7. Analyze each fraction by native electrophoresis and select the pure fractions. Combine and concentrate them by Mono Q chromatography to 5 mg/ml and store the concentrated pure protein at 23°C under argon or freeze as pellets in liquid N_2.

Ferredoxin Purification

1. Combine fractions from the first Q-Sepharose column with an A_{390}/A_{280} ratio of >0.15 (700 ml, 2800 mg) and apply them directly to a column (5 x 30 cm) of hydroxyapatite (high resolution) previously equilibrated with 50 mM Tris-HCl buffer (pH 8.3). Wash the column with the same buffer (300 ml) and elute the ferredoxin in 30-ml fractions with a gradient from 0 to 0.1 M potassium phosphate (1.2 liters) in the same buffer at 2 ml per minute. The ferredoxin starts to elute as soon as the gradient is applied.

2. Combine fractions from the hydroxyapatite column with an A_{390}/A_{280} ratio of >0.3 (330 ml, 1000 mg) and apply them to a column (2.5 x 25 cm) of DEAE-Sepharose Fast Flow preequilibrated with 50 mM Tris-HCl buffer (pH 7.8), using the FPLC system to dilute the fractions with an equivalent volume of the equilibration buffer as they are applied. Wash the column with equilibration buffer (400 ml) and elute the protein in 20-ml fractions at 2.5 ml per minute with sequential linear gradients from 0 to 0.15 M NaCl (90 ml) and 0.15 to 0.5 M NaCl (1.25 liters). The ferredoxin starts to elute as 0.29 M NaCl is applied.

3. Combine fractions with an A_{390}/A_{280} ratio of >0.42 (300 ml, 450 mg) and concentrate them to 10 ml by ultrafiltration. Apply the sample to a column (5 x 95 cm) of Sepharose CL-6B, previously equilibrated with 50 mM Tris-HCl buffer (pH 7.8), containing 0.1 M NaCl. Run the column at 3 ml per minute and collect fractions of 20 ml.

4. Combine fractions containing pure ferredoxin (A_{390}/A_{280} ratio = 0.56–0.58: 120 ml, 270 mg) and concentrate to 10 mg/ml using a column (HR 10/10) of Q-Sepharose preequilibrated with 50 mM Tris-HCl buffer (pH 8.0). Apply the sample directly to the column and elute at 0.8 ml per minute in 3-ml fractions with a gradient (60 ml) from 0 to 1.0 M NaCl. Store the protein at 23°C under argon or freeze as pellets in liquid N_2.

COMMENTARY

• All steps are performed under anaerobic conditions at 23°C. These procedures routinely yield approximately 30 mg of pure hydrogenase with a specific activity of 350 units/mg and 250 mg of pure ferredoxin. The final preparation of the hydrogenase gives rise to two protein bands after electrophoresis on polyacrylamide gels under nondenaturing conditions (Davis 1964). These appear to be electrophoretically distinguishable forms, possibly aggregates, of the same enzyme since both have the same subunit composition when analyzed by SDS-gel electrophoresis (Weber et al. 1972). The enzyme has an M_r of 180,000 by gel filtration and contains approximately 1 Ni and 30 Fe atoms per mole. The ferredoxin has an M_r of 7,500 (66 amino acids) and contains a single [4Fe-4S] cluster (Conover et al. 1990).

• The key to the purification of active hydrogenase is to carry out all purification procedures under strictly anaerobic and reducing conditions. The pure enzyme has a half-life in air of several hours. The ferredoxin is also routinely purified under strictly anaerobic conditions. Even though the pure ferredoxin shows no deterioration after many months in air at 23°C, anaerobic purification is recommended because some cluster degradation occurs during some of the chromatographic separations if these are carried out aerobically (Conover et al. 1990).

REFERENCES

Adams, M.W.W. 1990. The structure and mechanism of iron-hydrogenases. *Biochim. Biophys. Acta* **1020:** 115–145.

———. 1992. Novel iron sulfur clusters in metalloenzymes and redox proteins from extremely thermophilic bacteria. *Adv. Inorg. Chem.* **38:** 341–396.

———. 1993. Enzymes and proteins from organisms that grow near and above 100°C. *Annu. Rev. Microbiol.* **47:** 627–658.

———. 1994. Biochemical diversity among sulfur-dependent hyperthermophilic microorganisms. *FEMS Microbiol. Rev.* **15:** 261–277.

Aono, S., F.O. Bryant, and M.W.W. Adams. 1989. A novel and remarkably thermostable ferredoxin from the hyperthermophilic archaebacterium *Pyrococcus furiosus. J. Bacteriol.* **171:** 3433–3439.

Bryant, F.O. and M.W.W. Adams. 1989. Characterization of hydrogenase from the hyperthermophilic archaebacterium, *Pyrococcus furiosus. J. Biol. Chem.* **264:** 5070–5079.

Conover, R.C., A.T. Kowal, W. Fu, J.-B. Park, S. Aono, M.W.W. Adams, and M.K. Johnson. 1990.. Spectroscopic characterization of the novel iron-sulfur cluster in *Pyrococcus furiosus* ferredoxin. *J. Biol. Chem.* **265:** 8533–8541.

Davis, B.J. 1964. Disc electrophoresis. II. Method and application to human serum albumin. *Ann. N.Y. Acad. Sci.* **121:** 404–427.

Fiala, G. and K.O. Stetter. 1986. *Pyrococcus furiosus* sp. nov. represents a novel genus of marine heterotrophic archaebacteria growing optimally at 100°C. *Arch. Microbiol.* **145:** 56–61.

Lowry, O.H., N.J. Rosebrough, A.L. Farr, and R.J. Randall. 1951. Protein measurement with the Folin-phenol reagent. *J. Biol. Chem.* **193:** 265–275.

Weber, K., J.R. Pringle, and M. Osborn. 1972. Measurement of molecular weights by electrophoresis on SDS-acrylamide gels. *Methods Enzymol.* **26:** 2–27.

PROTOCOL 10

Archaeal Lipid Analysis

D.B. Hedrick and D.C. White

The Archaea are characterized by unique ether-linked isoprenoid lipids that do not occur in bacteria or eukarya (Tornebene and Langworthy 1979). The distinct archaeal lipid chemistry can be used for their detection in environmental samples or for the characterization of isolates. This procedure (Hedrick et al. 1991a) is an extension of a lipid extraction and analysis protocol for bacterial and eukaryotic lipids (Bligh and Dyer 1959; White et al. 1979). The overall approach is summarized in Figure 1. It is suitable for pure cultures and environmental samples and allows the separation and analysis of archaeal ether lipids

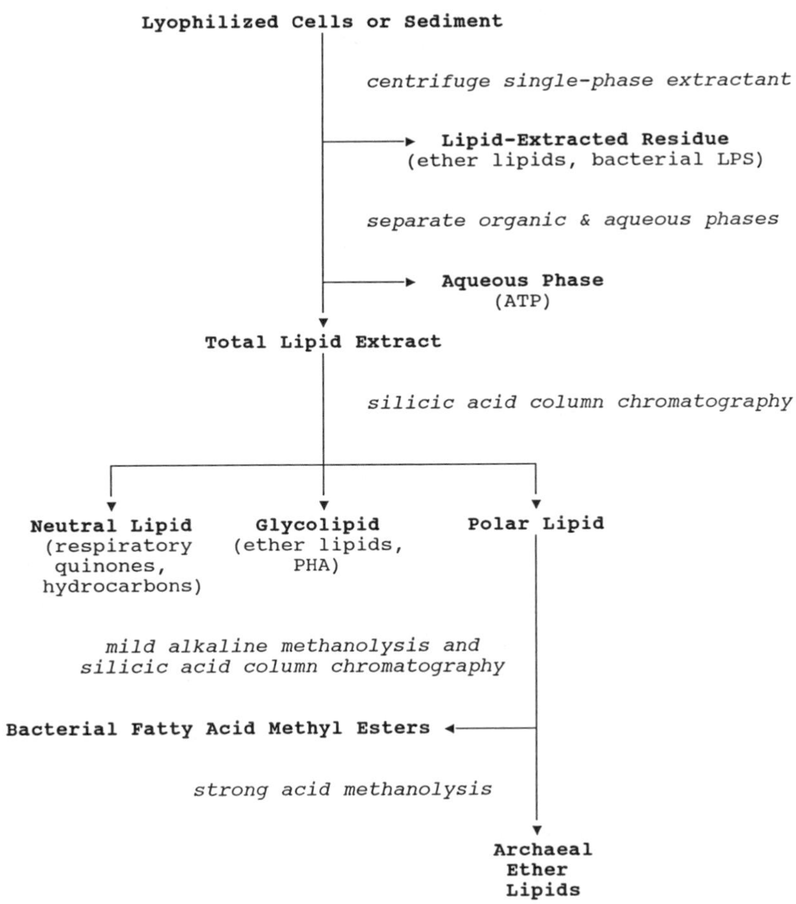

Figure 1 Lipid Extraction, Fractionation, and Derivatization Scheme (Guckert and White 1988).

and bacterial/eukaryotic ester-linked lipids on the same sample. Several other lipid classes are also made available for analysis by this protocol including hydrocarbons, eukaryotic triglycerides and sterols, and respiratory quinones.

MATERIALS

Disposable pasteur pipettes (9-inch; Baxter Diagnostics P5215-2)
Pyrex round-bottom flasks (250 ml; Baxter Diagnostics F4055-250)
Pyrex screw-capped test tubes (15 ml) with Teflon-lined caps (Baxter Diagnostics T1356-1A)
Separatory funnels (250 ml) with Teflon stopcocks (Baxter Diagnostics F7860-250)
Heating blocks (37°C and 100°C) for heating screw-cap test tubes
Tabletop centrifuge (Dyac II, Clay Adams 01103 or equivalent)
Vortex mixer (Scientific Industries 12-812 or equivalent)
Glass wool
Litmus paper
No. 2 Whatman folder filters (1202125) to filter lipid extract
Rotary Evaporator III (Büchi Labortechnik AK) with temperature-controlled water bath (37°C) to remove solvents from round-bottomed flasks
Detergent (Micro; Baxter Diagnostics C6286-6)
Dry nitrogen (industrial grade, available from welding suppliers) with temperature controlled water bath (37°C). Use a stream of dry nitrogen to remove solvents from test tubes.
Silicic acid (Unisil 100–200 mesh; Clarkson or equivalent). To bring to standard activity, heat silicic acid for at least 1 hour at 100°C in a *clean* oven. Use of the oven for other purposes, e.g., drying soils or plastics, will contaminate the silicic acid.
Solvents: chloroform, methanol, acetone, toluene, and hexane must be of the purest grade possible (Burdock and Jackson GC^2 grade from Baxter Diagnostics or equivalent). Concentrate an aliquot of each new lot by a factor of 100 and analyze by capillary gas chromatography to ensure that solvent is free of lipids.
Lipid-free H_2O for all dilutions (Barnstead Nanopure II or equivalent to deliver ≤17 ohm/cm and ≤20 ppb organic carbon). Extract overnight with and store over ~200 ml of chloroform per 4 liters of water. This will extract lipids and prevent microbial growth.
Formalin
Phosphate buffer (40 mM, pH 7.5): Dissolve 8.7 g of K_2HPO_4 in 1 liter of lipid-free H_2O and adjust pH to 7.4 with ~3.5 ml of 6 N HCl. Store over chloroform (see lipid-free H_2O).
Modified Bligh/Dyer single-phase extractant: Combine chloroform, methanol, and phosphate buffer to a volume ratio of 1:2:0.8.
Methanolic potassium hydroxide (0.2 M; prepare fresh before use). Rinse pellets with a small amount of hexane before weighing. Solution is hygroscopic and spoiled by small amounts of water.
Glacial acetic acid (1 N)
Toluene/methanol (1:1, v/v)
Hexane/chloroform (4:1, v/v)
Magic methanol: methanol/chloroform/concentrated hydrochloric acid (10:1:1, v/v). Make fresh daily.

Cautions: Chloroform is irritating to the skin, eyes, mucous membranes, and respiratory tract. Work in a chemical fume hood and wear gloves and safety glasses. Chloroform is a carcinogen and may damage the liver and kidneys.

Glacial acetic acid is volatile. Wear gloves and face protector and work in a chemical fume hood. Concentrated acids should be handled with great care.

Hydrochloric acid may be fatal if inhaled, swallowed, or absorbed through the skin. Extremely destructive to mucous membranes and upper respiratory tract, eyes, and skin. Wear gloves and safety glasses and work in a chemical fume hood.

Methanol is poisonous and can cause blindness if ingested in sufficient quantities. Adequate ventilation is necessary to limit exposure to vapors.

Toluene is irritating to the eyes, skin, mucous membranes and upper respiratory tract. Wear gloves and safety glasses and work in a chemical fume hood. Extremely flammable: Vapor may travel considerable distance to source of ignition and flash back.

METHODS

All glassware must be thoroughly washed with a good laboratory detergent, rinsed five times each with tap water and then with deionized H_2O, dried, wrapped in aluminum foil, and heated for 4 hours at 450°C. Disposable items (pasteur pipettes, glass wool) are not washed before heat treatment. Slow cooling will decrease the glassware's fragility. All plastic must be eliminated except for Teflon, e.g., Teflon stopcocks for separatory funnels and Teflon liners for test tubes. They are washed as above, solvent-rinsed with acetone, and air-dried. Samples must be either stored frozen at –20°C, fixed with buffered formalin, or lyophilized until extraction. Lyophilization allows dry weights of samples to be recorded.

Single-phase Extraction

1. Ideally, lyophilize and pulverize samples before extraction. For samples that must be analyzed wet, reduce the proportion of buffer to maintain the ratio of aqueous buffer to methanol and chloroform.

2. Combine sample and extractant and centrifuge in a test tube with a Teflon-lined screw-top lid.

3. Use a maximum of 1 mg of lyophilized cells or 1 g of lyophilized sediment per 1 ml of chloroform so as to not overload the extraction system. Proceed with extraction for 3 hours (up to 18 hours) with occasional mixing to ensure good contact between sample and extractant.

4. If components of extraction solvent are to be added separately (rather than premixed), add them in the order: methanol, buffer, and chloroform (mixing after each addition). Take care that the aqueous and organic phases do not separate into two phases.

5. Allow the mixture to settle. Tilt the container to observe whether there is free chloroform on the bottom.

6. If phase separation has occurred, add small amounts of methanol (1/20 original volume of methanol) until single-phase conditions are reestablished.

Separation of the Lipid-extracted Residue

1. Centrifuge to pellet the solids. Speed and time will depend on the container and sample matrix.
2. Decant the single-phase extractant into a separatory funnel or centrifugable container that holds at least 1.5 times the volume of extractant used. For very small sample volumes, use a screw-capped test tube. Filter the extract through lipid-clean glass wool if flocculent organics or clay refuse to pellet.
3. Air-dry the combined solids and any glass wool retentate overnight to yield lipid extracted residue. Reserve for analysis of residue-bound ether lipids.

Phase Separation

1. Add lipid-free H_2O and chloroform to the single-phase extract equal to the volume of chloroform used in the extraction solvent. Mix to ensure good contact, but not so vigorously that emulsions occur in high-biomass samples.
2. Allow the aqueous and organic phases to separate (~18 hours or until both phases are no longer cloudy).
3. Resolve emulsions by centrifugation until clear, addition of clean (450°C for 4 hours) sodium sulfate or sodium chloride, and/or storage at 0°C overnight.
4. Slowly drain the lower organic phase from the separatory funnel. Use a paper filter to trap solids and remove trace water.
5. If using a centrifugable container instead of a separatory funnel, transfer the upper aqueous phase to waste and then transfer the lower organic phase to a clean container. Take care not to transfer the aqueous phase with the organic phase.
6. Remove the organic solvent by rotary evaporation or with a stream of dry nitrogen at 37°C to yield the total lipid extract.
7. When dry, take up the total lipid in a small volume of chloroform. Cloudiness in this solution indicates the presence of water. Clear solution by adding chloroform/methanol (6:1) and a solvent removal to eliminate the water as the chloroform/methanol/water azeotrope. Repeat the chloroform test of total lipid until clear.
8. Store the solvent-free total lipid at –20°C under nitrogen.

Silicic Acid Column Chromatography

The total lipid is fractionated into neutral lipid, glycolipid, and polar lipid components by elution from silicic acid by chloroform, acetone, then methanol, respectively. Silicic acid columns are constructed from pasteur pipettes partially plugged with glass wool (heat-cleaned for 4 hours at 450°C after construction). Commercially available columns (i.e., Burdick and Jackson 7054G) may be used but each lot must be checked carefully for lipid contamination.

1. Silicic acid (0.5 g) is sufficient to fractionate lipid from 40 mg dry weight of cells. Increase the weight of silicic acid to prevent overloading the column.

2. Pack the column by wetting the glass wool with chloroform and transferring the silicic acid as a slurry in chloroform. Rinse silicic acid with chloroform. Tap the side of column to obtain a standard degree of packing of the silicic acid. The volumes of chloroform, acetone, and methanol are ten times the weight of silicic acid (0.5 g of silicic acid requires 5 ml of each solvent).

3. Dissolve the total lipid in a minimal volume of chloroform and *carefully* apply to the top of the column. Do not disturb the silicic acid surface.

4. Rinse the container that held the total lipid twice with small amounts of chloroform. Add each rinse to the column when the previous aliquot reaches the surface of the silicic acid.

5. Deduct the volume of chloroform used for the transfer from the volume of chloroform used to elute the neutral lipid fraction. Take care when adding solvents that the silicic acid bed does not go dry, causing channeling and void that degrade resolution.

6. When all the chloroform has been eluted, use a new test tube to add acetone to elute the glycolipid fraction. When the glycolipid fraction has eluted, use a new test tube to add methanol to elute the polar lipid fraction.

7. Remove the solvents with a stream of dry nitrogen and store the fractions at −20°C under nitrogen.

Mild Alkaline Methanolysis

The bacterial and eukaryotic ester-linked polar lipid fatty acids are transesterified to their methyl esters to facilitate their separation from the archaeal ether lipids.

1. Dissolve the dried glycolipid or polar lipid fraction in 1 ml of toluene/methanol and add 1 ml of methanolic KOH. Vortex the solution and incubate for 30 minutes at 37°C.

2. Cool the sample tubes, add 2 ml of hexane/chloroform, and mix.

3. Add 200 µl of 1 N acetic acid to neutralize the solution. Test neutrality with litmus paper.

4. Add 2 ml of lipid-free H_2O, mix with vortex mixer for at least 30 seconds, and centrifuge until a clear phase separation is achieved.

5. Transfer the upper organic phase to a clean test tube. Reextract the lower aqueous phase twice with 2 ml of hexane/chloroform.

6. Remove the solvent from the pooled extracts with a stream of dry nitrogen and store at −20°C under nitrogen.

Separation of Fatty Acid Methyl Esters from Ether Lipids

Mild alkaline methanolysis does not affect the glycolipid or polar lipid ethers, whereas the bacterial and eukaryotic ester-linked lipids are converted to fatty acid methyl esters.

1. Separate the ethers and fatty acids by a second silicic acid column performed exactly as above except omit the methanol step for glycolipid fractions and omit the acetone elution step for polar lipid fractions.

2. Recover the fatty acid methyl esters in the chloroform as neutral lipids and the ethers in the same fraction as in the first column chromatography step.

3. Remove the solvent and store the samples at –20°C under nitrogen.

Liberation of the Ether Lipids

The ether lipids of the lipid-extracted residue, glycolipid, and polar lipid fractions are cleaved with a strong acid methanolysis.

1. Add to the sample in a screw-topped test tube 1 ml of magic methanol and seal the tube. For the lipid-extracted residue of sediment samples, at least 10 ml of magic methanol per gram dry weight is required.

2. Examine the edge of the lip of the test tube and the Teflon liner of the cap to ensure smooth surfaces. Any small gap will allow the solvent to escape, spoiling the sample. Discard any test tube with a crack or a chip.

3. Incubate the sealed tubes in the heating block for 1 hour at 100°C.

4. After cooling, add 2 ml of hexane/chloroform and 2 ml of lipid-free H_2O, mix with vortex mixer for at least 30 seconds, and centrifuge at maximum speed for 5 minutes.

5. Transfer the upper organic phase to a clean test tube and reextract the lower aqueous phase with 2 ml of hexane/chloroform.

6. Remove the solvent from the pooled extracts with a stream of dry nitrogen at 37°C and store at –20°C under nitrogen.

COMMENTARY

- Although whole-cell (Welch 1991) and polar lipid (Guckert and White 1988) fatty acids have found wide application in detection, quantification, and identification of bacteria, archaeal lipids have not been so extensively utilized. One difficulty is that archaeal membrane lipids appear predominantly in the polar lipid, glycolipid, or lipid-extracted residue fraction, depending on the species (Hedrick et al. 1991a). For example, *Methanobacterium formicicum* and *Methanococcus maripaludis* had most of their membrane lipids in the lipid-extracted residue and glycolipid fraction, respectively. Rather than treating this unexpected variability as a problem, it was used as an opportunity to increase the information content of the archaeal lipid profile by analyzing all three fractions (Hedrick et al. 1991a).

• This procedure prepares the archaeal ether lipids of the lipid-extracted residue, glycolipid, and polar lipid fractions for analysis by supercritical fluid chromatography (Hedrick et al. 1991a), high-temperature gas chromatography (Nichols et al. 1993), or high-performance liquid chromatography (Demizu et al. 1992). Although it was designed to allow analysis of archaeal and bacterial lipids on the same sample, it is also useful for the elimination of fatty acid contaminants from archaeal pure cultures that arise from the complex media required by many methanogen and thermoacidophilic isolates.

• Many variations of this procedure are possible for specific applications. For example, the incorporation of $[^{14}C]$acetate into bacterial versus archaeal lipids may be monitored (Hedrick et al. 1991b) or their stable carbon isotope ratios determined (Rieley et al. 1991). Several purified fractions are prepared that are appropriate for analysis of specific classes of biomolecules. The magic methanol methanolysate of the lipid-extracted residue may also be analyzed by gas chromatography for the lipopolysaccharide fatty acids of gram-negative bacteria (Parker et al. 1982). The neutral lipid fraction contains the eukaryotic carbon storage compound triglyceride (Gehron and White 1982), the eukaryotic membrane sterols (White et al. 1980; Matcham et al. 1985), and the respiratory quinones of all three kingdoms (Collins and Jones 1981). The glycolipid fraction could alternatively be analyzed for the archaeal and bacterial carbon and energy storage compound poly-β-hydroxyalkanoate (Findlay and White 1987). This strategy can provide a great deal of information from a limited number of samples (Hedrick et al. 1992).

• A significant limitation of this method is the degradation of the recently described hydroxydiether lipids by the magic methanolysis step (Sprott et al. 1993). Development of a more gentle and reliable degradative procedure is being pursued in several laboratories.

• Another approach to the difficulty in extracting archaeal lipids from the cell solids has been taken by Nishihara and Koga (1987), who by making the chloroform/methanol extractant highly acidic, significantly increased the recovery of ether lipids. Since this damages the bacterial and eukaryotic ester-linked lipids, it is not appropriate for environmental studies involving both Archaea and either bacteria or eukaryotes.

REFERENCES

Bligh, E.G. and W.J. Dyer. 1959. A rapid method of total lipid extraction and purification. *Can. J. Biochem. Physiol.* **37:** 911–917.

Collins, M.D. and D. Jones. 1981. Distribution of quinone structural types in bacteria and their taxonomic implications. *Microbiol. Rev.* **45:** 316–354.

Demizu, K., S. Ohtsubo, S. Kohno, I. Miura, M. Nishihara, and Y. Koga. 1992. Quantitative determination of methanogenic cells based on analysis of ether-linked glycerolipids by high-performance liquid chromatography. *J. Ferment. Bioeng.* **73:** 135–139.

Findley, R.H. and D.C. White. 1987. A simplified method for bacterial nutritional status based on the simultaneous determination of phospholipid and endogenous storage lipid poly-β-hydroxybutyrate. *J. Microbial. Methods* **6:** 113–120.

Gehron, M.J. and D.C. White. 1982. Quantitative determination of the nutritional status of detrital microbiota and the grazing fauna by triglyceride glycerol analysis. *J. Exp. Mar. Biol. Ecol.* **64:** 145–158.

Guckert, J.B. and D.C. White. 1988. Phospholipid, ester-linked fatty acid analyses in microbial ecology. In *Proceedings of the Fourth International Symposium of Microbial Ecology*, pp. 455–459. Slovene Society for Microbiology, Ljubljana, Yugoslavia.

Hedrick, D.B., J.B. Guckert, and D.C. White. 1991a. Archaebacterial ether lipid diversity analyzed by supercritical fluid chromatography: Integration with a bacterial lipid protocol. *J. Lipid Res.* **32:** 659–666.

Hedrick, D.B., A. Vass, B. Richards, W. Jewell, J.B. Guckert, and D.C. White. 1991b. Starvation and overfeeding stress on microbial activities in high-solids high-yield methanogenic digesters. *Biomass Bioenergy* **1:** 75–82.

Hedrick, D.B., R.D. Pledger, D.C. White, and J.A. Baross. 1992. *In situ* microbial ecology of hydrothermal vent sediments. *FEMS Microbiol. Ecol.* **101:** 1–10.

Matcham, S.E., B.R. Jordan, and D.A. Wood. 1985. Estimation of fungal biomass in a solid substrate by three independent methods. *Appl. Microbiol. Biotechnol.* **21:** 108–112.

Nichols, P.D., P.M. Shaw, C.A. Mancuso, and P.D. Franzmann. 1993. Analysis of archaeal di- and tetraether phospholipids by high temperature capillary gas chromatography. *J. Microbial Methods* **18:** 1–9.

Nishihara, M. and Y. Koga. 1987. Extraction and composition of polar lipids from the archaebacterium, *Methanobacterium thermoautotrophicum. J. Biochem.* **101:** 997–1005.

Parker, J.H., G.A. Smith, H.L. Fredrickson, R.J. Vestal, and D.C. White. 1982. Sensitive assay, based on hydroxy fatty acids from lipopolysaccharide lipid A, for gram-negative bacteria in sediments. *Appl. Environ. Microbiol.* **44:** 1170–1177.

Rieley, G., R.J. Collier, D.M. Jones, G. Eglinton, P.A. Eakin, and A.E. Fallick. 1991. Sources of sedimentary lipids deduced from stable carbon-isotope analyses of individual compounds. *Nature* **352:** 425–427.

Sprott, G.D., C.J. DiCaire, C.G. Choquet, G.B. Patel, and I. Ekiel. 1993. Hydroxydiether lipid structures in *Methanosarcina* spp. and *Methanococcus voltae. Appl. Environ. Microbiol.* **59:** 912–914.

Tornebene, T.G. and T.A. Langworthy. 1979. Diphytanyl and dibiphytanyl glycerol ether lipids of methanogenic archaebacteria. *Science* **203:** 51–53.

Welch, D.F. 1991. Applications of cellular fatty acid analysis. *Clin. Microbiol. Rev.* **4:** 422–438.

White, D.C., R.J. Bobbie, J.S. Nickels, S.D. Fazio, and W.M. Davis. 1980. Nonselective biochemical methods for the determination of fungal mass and community structure in estuarine detrital microflora. *Bot. Mar.* **23:** 239–250.

White, D.C., W.M. Davis, J.S. Nickels, J.D. King, and R.J. Bobbie. 1979. Determination of the sedimentary microbial biomass by extractable lipid phosphate. *Oecologia* **40:** 51–62.

PROTOCOL 11

Glucose Metabolic Pathways in *Thermoplasma acidophilum* using ^{13}C- and ^{15}N-NMR Spectroscopy

K.J. Stevenson, L.A. Manning, M.J. Danson, and D. McIntyre

Nuclear magnetic resonance (NMR) spectroscopy permits the study of metabolism both in vivo and in vitro (Lundberg et al. 1990). For the study of systems in vivo, NMR spectroscopy offers the advantage of being noninvasive and is able to resolve, identify, and quantitate metabolites on a real-time basis. The study of cell-free systems by NMR spectroscopy is particularly useful when it is not possible to maintain the living organism in the NMR probe. The most useful nuclei to use in NMR studies include protons, carbon-13 (^{13}C), nitrogen-15 (^{15}N), and phosphorus-31 (^{31}P). Because the natural abundances of carbon-13 (1.1%) and nitrogen-15 (0.37%) are very low, it is common when studying these nuclei in vivo to improve the sensitivity by using a labeled precursor. The metabolic fate of the introduced labeled precursor can often be unambiguously determined.

This Protocol describes the metabolism of [^{13}C]glucose (uniformly labeled, C-1 labeled, and C-6 labeled) and [^{15}N]ammonium chloride in cell-free lysates of the moderately thermophilic Archaea *Thermoplasma acidophilum* (Danson and Hough 1992; Danson 1993). We used *Thermoplasma acidophilum* DSM strain 1728.

MATERIALS

Growth medium for *Thermoplasma acidophilum* (see Appendix 2, DSM 158 Medium).
Fernbach culture flasks (2800 ml; Corning No. 4420-2XL; VWR Scientific 29172-008)
Filters (0.22 μm pore size; Millipore GVWP 02500)
NMR spectrometers (wide-bore Bruker AM-400 and AMX-500)
Yeast extract
Glucose
DNase I (Boehringer Mannheim 104 132)
Glucose dehydrogenase (EC 1.1.1.119) (Boehringer Mannheim)
NAD^+, $NADP^+$, ATP (Boehringer Mannheim)
Magnesium sulfate (Fisher)
[^{13}C]glucose (uniformly labeled) [1-^{13}C]glucose, [6-^{13}C]glucose, and [^{15}N]ammonium chloride (Cambridge Isotope Laboratories)
Tris (2-amino-2-[hydroxymethyl]-1,3-propanediol) (Boehringer Mannheim 604 200)

NMR sample tubes for cell-free lysates (10 or 15 mm outer diameter) and NMR tubes for D_2O solutions of freeze-dried cell lysates and standard compounds (5 or 10 mm) (Wilmad Glass: 5 mm, 507-PP; 10 mm, 513-7PP; 15 mm, 515-7PP)
Deuterium oxide (D_2O; Wilmad Glass)
Sodium 2,2-dimethyl-2-silapentane-5-sulfonate (DSS; Wilmad Glass WG-233)
Tris-HCl buffer (200 mM, pH 8.0): 24.2 g Tris per liter adjusted to pH 8 with HCl

METHODS

Culture of *Thermoplasma acidophilum*

1. To culture *T. acidophilum*, use 2800-ml Fernbach flasks containing 1 liter of defined medium, yeast extract, trace elements, and glucose. Autoclave the salts, yeast extract, and glucose separately. Filter (0.22 μm) the solution of trace elements but do not autoclave.

2. Use a 5% inoculum in the *Thermoplasma* cultures in the prescribed mixture (see step 1) at 55°C with agitation at 120 rpm.

3. When the culture reaches an A_{650nm} of 0.4 (3 days), harvest the cells by centrifugation at 7000 rpm for 5 minutes at 25°C. Resuspend the cells at 25°C in fresh medium that lacks glucose and yeast extract and recover the cells by centrifugation.

Cell Lysis

1. Lyse the washed *Thermoplasma* cells for 30 minutes at 25°C by suspending them in 5 ml of 200 mM Tris-HCl buffer (pH 8.0) containing 1 μg of DNase I. DNase I reduces the viscosity of the lysate enabling easier mixing of additives and facilitates transfers to the NMR tubes.

2. Verify the activity of the cell lysate by assaying for glucose dehydrogenase activity ($NADP^+$).

NMR Spectrometry of Cell Lysates

1. Warm the cell-free lysate (7 ml) to 55°C in a 10-mm (or 15-mm) NMR tube and supplement with NAD^+, $NADP^+$, and ATP to a final concentration of 1 mM. Baseline NMR data were obtained using this solution. Introduce [^{13}C]glucose (uniformly labeled or specifically labeled at carbon-1 or carbon-6; 50 mg).

2. Supplement the cell lysate at 2-hour intervals over an 8-hour NMR monitor with 0.2-ml aliquots of a 100 mM cofactor solution (NAD^+, $NADP^+$, ATP) and 0.2-ml aliquots of a 100 mM $MgSO_4$ solution.

3. Transfer cell-free lysates of *Thermoplasma* (7 ml) in a similar manner in 200 mM Tris-HCl buffer (pH 8.0) to the NMR tube (10 or 15 mm) and add 50 mg of glucose and 5 mg of [^{15}N]H_4Cl.

4. After obtaining one-dimensional NMR data of the cell-free lysate, remove the solution and freeze-dry.

NMR Parameters

1. Record NMR spectra of the cell-free lysate system on a Bruker AM-400 wide-bore spectrometer operating in pulsed Fourier transform mode at 55°C.

2. Obtain ^{13}C-NMR spectra at a frequency of 100.6 MHz using a 15-mm switchable ^{13}C–^{31}P probe at 12-μsec pulses (45° flip angle) with broad-band composite pulse proton decoupling (WALTZ-16) and a recycle time of 2 seconds. Each spectrum consists of 600 scans (20 minutes each), and the spectra are calibrated by setting the signal from the β-anomeric carbon of glucose to 96.7 ppm.

3. Use the Bruker AM-400 spectrometer to record ^{15}N-NMR spectra at a frequency of 40.5 MHz in a 10-mm broad-band probe. Use the pulse length of 10 μsec (30° flip angle) with broad-band proton decoupling as above and a recycle time of 2 seconds.

4. Prior to transformation, set the first six points of all ^{15}N-free induction decays to zero in order to eliminate distortions caused by probe ringing and pulse breakthrough. This leads to an effective deadtime of 170 μsec. The spectra are referenced to external nitrate (0 ppm).

5. Subject both ^{13}C- and ^{15}N-free induction decays to a Lorentz line broadening of 10 Hz prior to transformation.

NMR Spectroscopy of Freeze-dried Lysates

1. Record NMR spectra of freeze-dried lysates on a Bruker AMX-500 spectrometer operating at 500.13 MHz for ^{1}H, 125.77 MHz for ^{13}C, and 50.69 MHz for ^{15}N.

2. Obtain spectra at 25°C in a 5-mm triple resonance ^{1}H, ^{13}C, ^{15}N probe. The 90° pulse lengths are 8 μsec for ^{1}H, 10 μsec for ^{13}C, and 23 μsec for ^{15}N.

3. Obtain one-dimensional proton spectra in D_2O solution with presaturation of the residual monodeuterated water (HOD) resonance at 4.8 ppm and reference to internal DSS (0 ppm). One-dimensional ^{13}C spectra are acquired with composite pulse proton decoupling.

4. Use standard Bruker pulse programs to obtain two-dimensional heteronuclear multiple quantum coherence (HMQC), heteronuclear multiple bond coherence (HMBC), and double-quantum-filtered correlation spectroscopy (COSY) NMR spectra in D_2O solutions.

Identification of Metabolites

1. Assign spectra of the cell-free lysates through the introduction of standard compounds (where available) to the solution. Alternatively, to identify the metabolites, analyze the two-dimensional spectra and compare with known chemical shifts of the metabolic compounds.

2. Through experiments using uniformly labeled [^{13}C]glucose, it is possible to distinquish resonances arising from this precursor because homonuclear ^{13}C coupling is observed in ^{13}C spectra. In all cases, determine the fate of

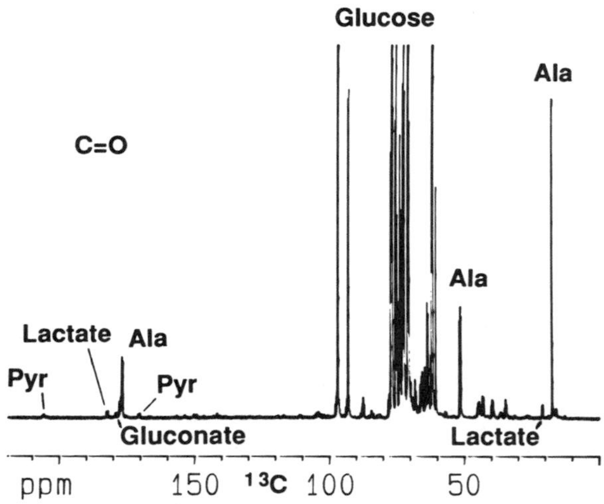

Figure 1 Carbon-13 NMR Spectrum (Proton Decoupled) Obtained after Addition of Glucose Uniformly Labeled with Carbon-13. Signals showing fine structure arising from carbon-13 coupling represent compounds derived from the carbon-13-labeled glucose.

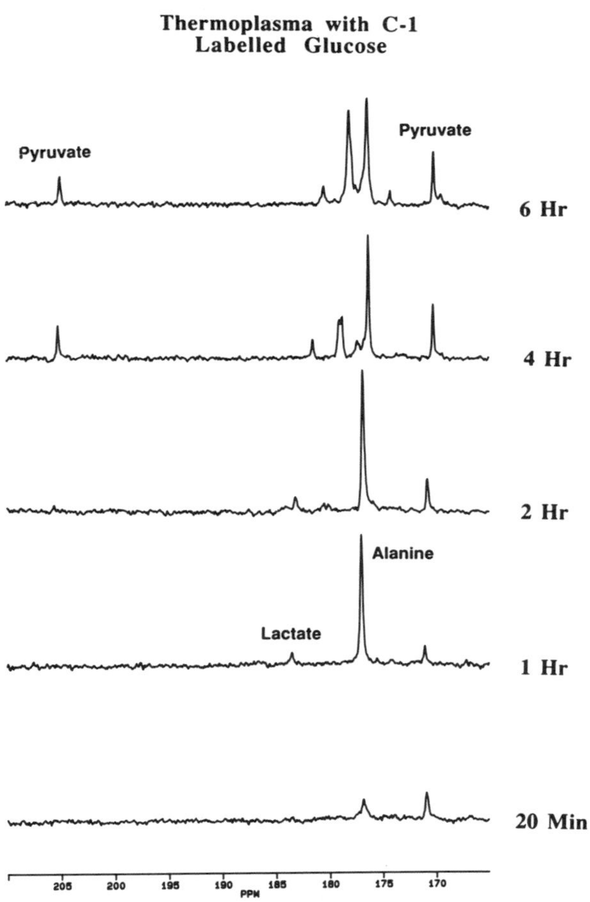

Figure 2 Carbon-13 NMR Spectra (Carbonyl Region) Obtained after Addition of Glucose Uniformly Labeled with Carbon-13. Signals were collected continuously and are shown at the times specified. The times of appearance of specific compounds are noted.

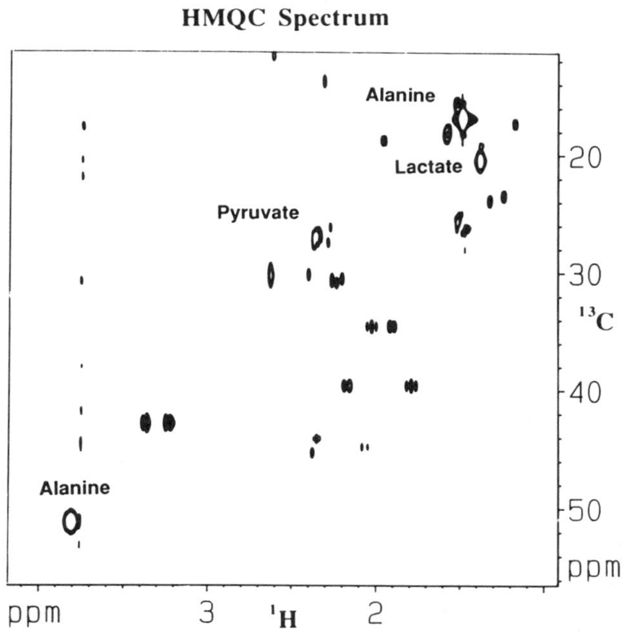

Figure 3 Carbon-Proton HMQC NMR Spectrum Obtained after Addition of Glucose Uniformly Labeled with Carbon-13. This spectrum correlates carbon-13 chemical shifts with those of directly attached protons.

the ^{13}C label by the presence of heteronuclear ^{1}H–^{13}C coupling in one- and two-dimensional proton NMR spectra of the freeze-dried lysate.

COMMENTARY

- ^{13}C-NMR spectra (proton decoupled), obtained after the addition of uniformly labeled [^{13}C]glucose, revealed signals showing the fine structure arising from ^{13}C couplings of compounds into which the ^{13}C label had been incorporated (see Figs. 1 and 2). Gluconate, pyruvate, alanine, and lactate were observed and their identity was confirmed by two-dimensional HMQC (Fig. 3) and HMBC NMR. The HMBC NMR spectra correlated carbonyl carbons with protons on adjacent carbons and were used to assign the carbonyl signals in the one-dimensional ^{13}C-NMR spectra. The HMQC NMR spectrum correlated ^{13}C chemical shifts with those of directly attached protons (Fig. 3).

- Other metabolites known to comprise the nonphosphorylated Entner-Doudoroff pathway (De Rosa et al. 1984; Danson and Hough 1992; Danson 1993) were not confirmed likely because of their low concentrations. A key intermediate in this pathway, 2-keto-3-deoxygluconate, could not be verified because of the lack of an authentic standard. However, signals arising at 185 ppm in the ^{13}C-NMR spectra with uniformly labeled [^{13}C]glucose were tentatively assigned to 2-keto-3-deoxygluconate. NMR spectra of lysates of *Thermoplasma acidophilum* metabolizing [^{13}C]glucose reaffirmed the labeling of gluconate and pyruvate; 2-keto-3-deoxygluconate at 185 ppm was tentatively assigned. NMR spectra of these cell-free lysates metabolizing [^{13}C]glucose showed the ^{13}C label in gluconate (2-keto-3-deoxygluconate tentative) and pyruvate. Arising from pyruvate, the end-products of the modified Entner-Doudoroff pathway were lactate and alanine.

• The ^{15}N-NMR spectra of the cell lysate containing $[^{15}N]H_4Cl$ and (unlabeled) glucose confirmed the formation of ^{15}N-labeled alanine, proline, and glutamine.

• The ^{13}C- and ^{15}N-NMR spectroscopy studies support the existence of the modified Entner-Doudoroff pathway in *Thermoplasma acidophilum* (De Rosa et al. 1984; Danson and Hough 1992; Danson 1993) as reported in the hyperthermophile *Pyrococcus furiosus* (Mukund and Adams 1991; Schäfer and Schönheit 1992).

REFERENCES

Danson, M.J. 1993. Central metabolism in the Archaea. *New Compr. Biochem.* **26:** 1–24.

Danson, M.J. and D.W. Hough. 1992. The enzymology of archaebacterial pathways of central metabolism. *Biochem. Soc. Symp.* **58:** 7–21.

De Rosa, M., A. Gambocorta, B. Nicolaus, P. Giardina, E. Poerio, and V. Buonocore. 1984. Glucose metabolism in the extreme thermophilic archaebacterium *Sulfolobus solfataricus. Biochem. J.* **224:** 407–414.

Lundberg, P., E. Harmsen, C. Ho, and H.J. Vogel. 1990. Nuclear magnetic resonance studies of cellular metabolism. *Anal. Biochem.* **191:** 193–222.

Mukund, S. and M.W.W. Adams. 1991. The novel tungsten-iron-sulfur protein of the hyperthermophilic archaebacterium *Pyrococcus furiosus*, is an aldehyde ferredoxin oxidoreductase. *J. Biol. Chem.* **266:** 14208–14216.

Schäfer, T. and P. Schönheit. 1992. Maltose fermentation to acetate, CO_2 and H_2 in the anaerobic hyperthermophilic archeon *Pyrococcus furiosus.* Evidence for a novel sugar fermentation pathway. *Arch. Microbiol.* **158:** 188–202.

PROTOCOL 12

Purification of Plasmids from Thermophilic and Hyperthermophilic Archaea

F. Charbonnier, P. Forterre, G. Erauso, and D. Prieur

Presented here is a modification of the sodium dodecyl sulfate (SDS) lysis method described by Maniatis et al. (1982). Our method involves three steps: harvesting and lysis of the cells, isolation of total DNA, and purification of plasmid DNA. The principle is based on complete cell lysis, followed by drastic deproteinization, efficient recovery of total DNA, and finally the separation of the genomic and plasmid DNAs. In our laboratory, this method has been used to purify plasmids of low-copy number, as in case of pSSV1, from the non-irradiated *Sulfolobus shibatae* cells (Forterre et al. 1992). This procedure can also be used for other cases, such as that of the pGT5 plasmid isolated from the hyperthermophilic strain *Pyrococcus abysii* (formerly GE5) (Erauso et al. 1992) or pSL10 from *Desulfurolobus ambivalens* (Charbonnier and Forterre 1994). This purification technique includes applications that necessitate the analysis of all plasmid molecules, for example, for studies on DNA topology (Charbonnier et al. 1992). The DNA is suitable for all common applications, such as restriction analysis, cloning, transformation, and sequencing.

MATERIALS

Corex tubes (30 and 15 ml, PolyLabo 11048 and 11047)
Centrifuge (J2-MC; Beckman)
Ultracentrifuge (TL-100; Beckman)
Rotors (JS-13, VTI-65, TLA100-2; Beckman)
Quick-seal centrifuge tubes (5.5 and 1.5 ml, Beckman 342.883 and 344.625)
Teflon tubes (50 ml; BioBlock K99431)
Long-wave UV light source
Ethanol (70%, 95%)
1-butanol (saturated with 5 M NaCl; Sigma B 7906)
Cesium chloride (Sigma C 3032)
Ethidium bromide (10 mg/ml in TE solution; Sigma E 1510)
Chloroform/isoamyl alcohol (24:1; Sigma I 1885)
N-lauroyl sarcosine (sodium salts; Sigma L 5125)
Phenol (AquaPhenol, Appligene 130.181)

Proteinase K (Boehringer Mannheim 745.723)
RNase (DNase-free; Boehringer Mannheim 1119.915)
SDS (sodium dodecyl sulfate; Boehringer Mannheim 1028.685)
TNE solution: 100 mM Tris-HCl, 100 mM NaCl, 50 mM EDTA (pH 7.5)
TE solution: 10 mM Tris-HCl, 2 mM EDTA (pH 7.5)
10 mg/ml RNase (DNase-free, 1 ml): 10 mg of DNase-free RNase and 1 ml of TE solution (pH 7.5). Incubate for 5 minutes at 100°C, cool to room temperature, and freeze aliquots at –20°C.

Cautions: Ethidium bromide is a powerful mutagen and moderately toxic. Wear gloves when working with solutions that contain this dye. After use, decontaminate and discard according to local safety office regulations.

Phenol is highly corrosive and can cause severe burns. Wear gloves, protective clothing, and safety glasses and work in a chemical fume hood. Any areas of skin that come in contact with phenol should be rinsed with a large volume of water or PEG 400 and washed with soap and water; do not use ethanol!

Wear a mask when weighing SDS.

Ultraviolet radiation is dangerous, particularly to the eyes. To minimize exposure, make sure that the ultraviolet light source is adequately shielded and wear protective goggles or a full safety mask that efficiently blocks ultraviolet light.

METHODS

Step 1: Harvesting and Lysis of the Archaeon

1. Harvest the archaeal cells from 1 liter of late log phase culture by centrifugation at 6000*g* for 15 minutes at room temperature.

2. Resuspend cell pellets in a total volume of 8 ml of TNE solution (pH 7.5) at room temperature. Homogenize suspension.

3. Transfer the solution into a 30-ml Corex tube and add 1 ml of 10% *N*-lauroyl sarcosine. Invert slowly several times at room temperature.

4. Add 1 ml of 10% SDS at room temperature and invert tube gently.

5. Add 0.5 ml of 20 mg/ml proteinase K and incubate for 3 hours at 50°C. After incubation, the lysate must be translucent and viscous.

Step 2: Isolation of the Total DNA

1. Transfer the viscous lysate into a 50-ml Teflon tube. Add 11 ml of TE-saturated phenol (pH 7.5) and stir for 10 minutes. Spin at 5400*g* for 10 minutes at 20°C. Transfer 90% or more of the aqueous layer into a 50-ml Teflon tube. Repeat twice.

2. Add 11 ml of chloroform/isoamyl alcohol (24:1) and stir for 10 minutes. Spin at 5400*g* for 10 minutes at 20°C. Transfer 90% or more of the aqueous solution into three 15-ml Corex tubes, containing 3 ml in each tube.

3. Add 8 ml of 95% ethanol in each tube and incubate for 1 hour at –20°C.

4. Centrifuge at 15,000*g* for 30 minutes at 4°C (e.g., use a swingout rotor such as a Beckman JS-13 rotor).

5. Resuspend pellets in 3 ml of 70% ethanol and transfer all of them into a 15-ml Corex tube. Centrifuge at 12,000*g* for 10 minutes at 20°C. The pellet must be washed again once with 70% ethanol and centrifuged.

6. Air-dry the pellet for 2 or 3 minutes.

7. Completely resuspend the pellet in 2 ml of TE solution (pH 7.5). Add 3 µl of 10 mg/ml DNase-free RNase and incubate on a rotator for 1 hour at 37°C.

Step 3: Purification of the Plasmid DNA

1. Resuspend cell pellets in a total volume of 4.5 ml of TE solution (pH 7.5). Add 4.8 g of cesium chloride and 0.2 ml of a 10 mg/ml ethidium bromide solution in TE. Mix gently and transfer to a 5.5-ml Beckman Quick-seal centrifuge tube. Spin at 45,000 rpm in a Beckman VTI-65 rotor for 16 hours at 17°C.

2. Visualize the plasmid band with long-wave UV light (365 nm). Remove plasmid DNA, extract ethidium bromide with 5 M NaCl saturated 1-butanol, and dialyze several successive times against 1 liter of 10^{-2} mM TE solution (pH 7.5) at 4°C.

Alternative Protocol

Before the third purification step, load an aliquot of the preparation on an agarose gel. Run and subsequently stain the gel with a 2 µg/ml ethidium bromide solution for 20 minutes. After several washes in double-distilled H_2O, illuminate the gel under UV light at 254 nm. If the supercoiled form I of the plasmid is not visible, use the following alternate final purification step.

1. Resuspend cell pellets in a total volume of 3.1 ml of TE solution (pH 7.5). Add 3.52 g of cesium chloride and 0.12 ml of 10 mg/ml ethidium bromide in TE. Mix and transfer in two 1.5-ml Beckman Quick-seal centrifuge tubes. Spin at 100,000 rpm in a Beckman TLA100-2 rotor at 17°C for 16 hours in the TL-100 Beckman centrifuge.

2. Proceed as described in Step 3.2. The total volume of the gradient suspension and the tube diameter are lower than in the basic protocol described above in Step 3.1. This procedure permits increased recovery of the form I plasmid band.

COMMENTARY

- We originally tried to purify plasmids from thermophilic and hyperthermophilic Archaea using the alkaline lysis method (Maniatis et al. 1982). In the case of plasmids from *Sulfolobus shibatae*, the new hyperthermophilic archaeon *Pyrococcus abysii* (formerly GE5), and *Desulfurolobus ambivalens*, all the plasmid DNA was dramatically lost or fragmented. We then decided to isolate the total DNA with minimal pH perturbations as described above. This method does not seem to be influenced by the growth medium or the growth phase of the culture at the time of harvesting. The more important steps are Steps 1.2 and 1.5. The suspension of the cells must be complete and homogeneous in order to maximize the number of lysed cells. The protease treatment for 3 hours at 50°C is essential in order to disrupt all the DNA-protein interactions. If this is not the case, these complexes will be lost at the interface between the two layers during the phenol treatment.

- We also used this method to prepare plasmids from thermophilic bacteria, such as *Bacillus coagulans* or *Thermus thermophilus* HB8 (Charbonnier and Forterre 1994). In these cases, Step 1.2 was modified by a resuspension in a total volume of 8 ml of TNE solution (pH 7.5) containing 8 mg of lysozyme, followed by an incubation for 1 hour at 37°C. Although we have not yet investigated all branches of thermophilic and hyperthermophilic Archaea, this method has generally been successful.

REFERENCES

Charbonnier, F. and P. Forterre. 1994. Comparison of plasmid DNA topology among mesophilic and thermophilic eubacteria and archaebacteria. *J. Bacteriol.* **176:** 1251–1269.

Charbonnier, F., G. Erauso, T. Barbeyron, D. Prieur, and P. Forterre. 1992. Evidence that a plasmid from a hyperthermophilic archaebacterium is relaxed at physiological temperatures. *J. Bacteriol.* **174:** 6103–6108.

Erauso, G., F. Charbonnier, T. Barbeyron, P. Forterre, and D. Prieur. 1992. Preliminary characterization of a hyperthermophilic archaebacterium with a plasmid, isolated from a North Fiji basin hydrothermal vent. *C.R. Acad. Sci.* **314:** 387–393.

Forterre, P., F. Charbonnier, E. Marguet, F. Harper, and G. Henckes. 1992. Chromosome structure and DNA topology in extremely thermophilic archaebacteria. *Biochem. Soc. Symp.* **58:** 99–112.

Maniatis, T., E.F. Fritsch, and J. Sambrook. 1982. *Molecular cloning: A laboratory manual.* Cold Spring Harbor Laboratory, Cold Spring Harbor, New York.

PROTOCOL 13

Transfection of *Sulfolobus solfataricus*

C. Schleper and W. Zillig

Phage SSV1, originally isolated from *Sulfolobus shibatae* (Martin et al. 1984), has been shown to infect *S. solfataricus* and thus a plaque assay was established for this virus/host system (Schleper et al. 1992). The 15.5-kbp circular, double-stranded DNA of the phage (Palm et al. 1991) has been used in transfection experiments to find and optimize conditions for the transformation of *Sulfolobus* by exogenous DNA (Schleper et al. 1992). Transfectants have been scored by plaque formation. Described here is a method for DNA uptake yielding a transfection efficiency of 10^6 transfectants per microgram of DNA, with a transfection frequency of up to 10^{-4} (transfectants per surviving cell), such that its use for genetic complementation and for the development of transformation vectors for *Sulfolobus* appears to be possible.

Sulfolobus shibatae B12 (DSM 5389) is lysogenic for phage SSV1 and was used as a source for preparing phage DNA. *S. solfataricus* P1 (DSM 1616) and P2 (DSM 1617) are transfectable with phage DNA. All organisms were grown heterotrophically under aerobic conditions in Brock's medium (Brock et al. 1972) supplemented with carbon sources.

MATERIALS

Liquid medium for *Sulfolobus* (see Appendix 2, DSM 88 Medium)

SSVI DNA. Prepare from lysogenic cells of *S. shibatae* (or *S. solfataricus*) or from phage particles obtained after UV-irradiation of a lysogen (Martin et al. 1984) using CsCl density gradient centrifugation (Yeats et al. 1982).

Gelrite (6.5 g/liter; Merck, Kelco Division 58-46-3070) to solidify medium for plates plus 2 g of $MgCl_2{\cdot}6H_2O$ and 440 mg of $CaCl_2{\cdot}2H_2O$ (Grogan 1989)

High-voltage electroporation device (Bio-Rad gene-pulser 165-2077 and pulse controller 165-2098)

Cuvettes with an electrode gap of 1 mm (Bio-Rad 165-2089)

Centrifuge (Heraeus Instruments Minifuge) with a swinging-bucket rotor

Sucrose (20 mM) dissolved in double-distilled H_2O (natural pH of 5.6–5.8)

Caution: Ultraviolet radiation is dangerous, particularly to the eyes. To minimize exposure, make sure that the ultraviolet light source is adequately shielded and wear protective goggles or a full safety mask that efficiently blocks ultraviolet light.

METHODS

1. Grow a 50-ml *S. solfataricus* culture (inoculated with 1 ml of a late log phase starter culture) on a shaker at 80°C to an optical density of 0.1–0.3 A_{600} (~15 hours). Grow another overnight culture (200 ml) for the indicator lawn.

2. After cooling the culture (50 ml) on ice, pellet the cells by low-speed centrifugation (3000 rpm in a swinging-bucket rotor) and discard the supernatant completely. If not otherwise indicated, perform all of the following steps at 4°C.

3. Resuspend the cells in an equal volume of 20 mM sucrose and centrifuge again. To decrease the amount of residual salts from the growth medium, wash the cells twice more with 20 mM sucrose, first in 1/2 and then in 1/50 of the original culture volume. Finally, resuspend the pellet in 300–500 μl of 20 mM sucrose (yielding a concentration of ~10^{10} cells/ml). *Note:* Cells are 100% viable up to 2 hours when stored on ice.

4. Mix 50 μl of the cell suspension with 1 μl of SSV1 DNA (typically at concentrations of 50–300 ng/ml), transfer the mixture to an electroporation cuvette, and cool the cuvette on ice.

5. Electroshock cells with a gene pulser apparatus at 15 kV/cm, 400 Ω, 25 μF that yields time constants of 9.0–9.5 msec. After electroporation, immediately dilute the cells with 1 ml of growth medium (containing sucrose and yeast extract) and incubate for 1 hour at 80°C.

6. Remove aliquots for serial dilutions and mix each with 500 μl of tenfold concentrated mid-log cells of *S. solfataricus* (indicator lawn). Preheat the mixture to 60°C in a glass tube and add 3 ml of freshly prepared hot (70–80°C) soft overlay medium containing sucrose and yeast extract as carbon sources and 0.35% Gelrite. Pour the overlay onto Gelrite plates containing the same growth medium and incubate the plates at 80°C upside down under a moist atmosphere in a jar. *Note:* Turbid plaques with a diameter of 3–6 mm are visible after 2 days.

7. To check the survival rate after electroporation, plate appropriate dilutions on tryptone plates and compare to a nonelectroporated control. Cells can simply be plated with a spreader or alternatively be poured into a 1–3-ml soft layer, which results in a plating efficiency of nearly 100%. *Note:* Colonies appear within 5–7 days. Cell survival should be 30–70% of the original viable count.

COMMENTARY

- Note that the medium used in the plaque assay is different from that used in the plating of colonies. This is due to the fact that reproducible plating of individual colonies has only been observed on tryptone plates, whereas plaque formation can best be monitored on sucrose/yeast plates.

- DNAs that contain a considerable amount of residual salts will reduce the time constant in the electroporation procedure, which results in a lower cell survival and reduced transfection efficiency.

• Transfection of *S. solfataricus* with SSV1 DNA prepared from *S. shibatae* did not show a lower efficiency than with DNA preparations from lysogens of *S. solfataricus*, suggesting that there is no restriction or modification system in *S. shibatae*.

• Transfection of *S. solfataricus* has been demonstrated with linearized SSV1 DNA (cut at the unique *Bam*HI and *Xho*I sites), as well as transfection with SSV1 DNA that has been shuttled through *Escherichia coli* DH5α (SSV1 DNA was cut from a pBR322:SSV1 plasmid with *Bam*HI and religated after dilution; C. Schleper, unpubl.).

REFERENCES

Brock, T.D., K.M. Brock, R.T. Belly, and R.L. Weiss. 1972. *Sulfolobus*: A new genus of sulfur-oxidizing bacteria living at low pH and high temperature. *Arch. Microbiol.* **84:** 54–68.

Grogan, D. 1989. Phenotypic characterization of the archaebacterial genus *Sulfolobus*: Comparison of five wild-type strains. *J. Bacteriol.* **171:** 6710–6719.

Martin, A., S. Yeats, D. Janekovic, W.D. Reiter, W. Aicher, and W. Zillig. 1984. SAV1, a temperate u.v.-inducible DNA virus-like particle from the archaebacterium *Sulfolobus acidocaldarius* isolate B12. *EMBO J.* **3:** 2165–2168.

Palm, P., C. Schleper, B. Grampp, S. Yeats, P. McWilliam, W.D. Reiter, and W. Zillig. 1991. Complete nucleotide sequence of the virus SSV1 of the archaebacterium *Sulfolobus shibatae*. *Virology* **185:** 242–250.

Schleper, C., K. Kubo, and W. Zillig. 1992. The particle SSV1 from the extremely thermophilic archaeon *Sulfolobus* is a virus: Demonstration of infectivity and of transfection with viral DNA. *Proc. Natl. Acad. Sci.* **89:** 7645–7649.

Yeats, S., P. McWilliam, and W. Zillig. 1982. A plasmid in the archaebacterium *Sulfolobus acidocaldarius*. *EMBO J.* **1:** 1035–1038.

PROTOCOL 14

Preparation of Genomic DNA from Sulfur-dependent Hyperthermophilic Archaea

V. Ramakrishnan and M.W.W. Adams

Preparation of unsheared genomic DNA from elemental sulfur-reducing hyperthermophilic Archaea has been a major problem in our laboratory using protocols that involve a phenol extraction step. Described here is a rapid and simple nonphenol procedure for the preparation of unsheared genomic DNA from sulfur-reducing hyperthermophiles.

MATERIALS

Tabletop centrifuge (International Equipment, Centra-4B)
Microcentrifuge (Eppendorf 5415C)
Microfuge tubes (1.5 ml)
Vacuum centrifuge (Savant Speed Vac SC100)
Transparent centrifuge tubes (15 ml polystyrene, Becton Dickinson 2099)
Pasteur pipettes (hooked; Fisher 13-678-20D)
Ethanol (95%)
Salt-out mixture (BIO 101 7104)
Cell suspension solution: 0.5 M Tris-Cl (pH 8.0), 0.2 M EDTA, 0.46 M NaOH
RNase mixture: RNase A (1 mg/ml; Boehringer Mannheim 109 142) in 50% glycerol, 0.02 M NaCl, 0.001 M EDTA, 0.05 M Tris-Cl (pH 7.5–7.6)
Cell lysis/denaturing solution: 15% SDS (sodium dodecyl sulfate) (pH 6.6)
Protease mixture: proteinase K (20 mg/ml) in 0.05 M Tris-Cl (pH 8.0) containing 0.005 M $CaCl_2$
TE Buffer: 0.01 M Tris-Cl (pH 8.0), 0.001 M EDTA

METHODS

1. Aerobically transfer a 20-ml culture (see Protocols 1 and 2) of an S^o-dependent hyperthermophile grown to log phase (typically $A_{600} = 0.6$) to a transparent centrifuge tube. Spin the cells in a tabletop centrifuge at 5000 rpm for 7 minutes.

2. Discard the supernatant and add 1.85 ml of cell suspension solution; mix until the solution is homogeneous.

3. Add 50 μl of RNase mixture and mix thoroughly. Add 100 μl of cell lysis solution and mix well. Incubate for 15 minutes at 55°C.

4. Add 25 μl of protease mixture and mix thoroughly; incubate for 60 minutes at 55°C.

5. Add 500 μl of Salt-out mixture. Mix the solution thoroughly but gently. Divide the sample into 1.5-ml microfuge tubes and refrigerate for 10 minutes at 4°C.

6. Spin for 10 minutes at maximum speed in a microcentrifuge. Carefully collect the supernatant and transfer to a 10-ml transparent centrifuge tube (if a precipitate remains in the supernatant, spin down again until it is clear).

7. To the supernatant, add 2 ml of TE buffer and 8 ml of ethanol. Allow to stand for 2 minutes and then gently invert the tube several times. DNA threads or filaments will appear in the solution.

8. Scoop out the DNA using a hooked pasteur pipette. Alternatively, spin down the filaments at 5000 rpm for 15 minutes (the white genomic DNA pellet should be visible at the bottom of the tube). Discard the supernatant and vacuum dry for approximately 20 minutes to remove the excess ethanol.

9. Resuspend the pellet overnight in 100 μl of TE buffer. Run it on a 0.7% agarose gel to check for purity and to concentrate the sample.

10. If the DNA concentration is very low, add 0.1 volume of 3 M sodium acetate and 3 volumes of ethanol. Mix the solution slowly. A white pellet should be visible. Continue mixing until the pellet can be recovered by scooping out with a hooked pasteur pipette. Air-dry for a few minutes and suspend in appropriate volume of TE buffer.

COMMENTARY

• This procedure routinely provides sufficient quantities (≥2 μg) of unsheared genomic DNA from 20-ml cultures of various hyperthermophiles, including from *Pyrococcus furiosus*, *Thermococcus litoralis*, *Pyrococcus* strain ES4, and *Thermococcus* strain ES1. It is a simple and rapid procedure and does not need any special precautions such as removal of elemental sulfur from cell pellets and refrigerated centrifugation.

• We have observed significant fragmentation of DNA prepared by phenol-based extraction procedures, which presumably arises from the generation of radicals mediated by phenol in the presence of iron sulfides. Other non-phenol extraction procedures besides the one described here will likely work equally well.

REFERENCES

Ausubel, F.M., R. Brent, R.E. Kingston, D.D. Moore, J.G. Seidman, J.A. Smith and K. Struhl, eds. 1990. *Current protocols in molecular biology.* John Wiley, New York.

PROTOCOL 15

RNA Extraction from Sulfur-utilizing Thermophilic Archaea

J. DiRuggiero and F.T. Robb

The availability of highly purified RNA is crucial to the success of many protocols that apply to the molecular biology of Archaea, for example, cDNA cloning and the analysis of transcription. We present here a method for extracting RNA from hyperthermophiles.

MATERIALS

Centrifuge (Beckman J2-21) with a JA 20 rotor (Beckman 334831)
Corex tubes (15 ml; Corning 8441-15)
Vacuum centrifuge (Speed Vac concentrator SVC 100H Savant)
Guanidine thiocyanate (USB 32815)
Phenol (TE-saturated, pH 4.6; Amresco 0981)
Sodium acetate (2 M, pH 4.0)
Chloroform/isoamyl alcohol (24:1, v/v)
Phenol/chloroform (1:1, v/v)
Ethanol (70%, 100%, v/v)
Sodium acetate (3 M, pH 5.25)
DEPC (diethyl pyrocarbonate; Sigma D 5758)
Ethidium bromide
TE buffer (1x): 0.01 M Tris-HCl, 0.1 mM EDTA (pH 7.6)
Extraction Buffer: 250 ml of guanidine thiocyanate, 17.6 ml of sodium citrate (0.75 M, pH 7.0), 26.4 ml of sarcosyl (65°C), 293 ml of water (65°C). After passing through a 0.22-μm polyethersulfone filter (Gelman 09730-26), the solution can be stored several months at room temperature. Before use, add β-mercaptoethanol (Amresco 0482) to the stock solution to a final concentration of 0.75% (v/v).

Caution: β-Mercaptoethanol may be fatal if inhaled or absorbed through the skin and is harmful if swallowed. High concentrations are extremely destructive to the mucous membranes, the upper respiratory tract, the skin, and the eyes. Wear gloves and safety glasses and work in a chemical fume hood.

Chloroform is irritating to the skin, eyes, mucous membranes, and respiratory tract. Work in a chemical fume hood and wear gloves and safety glasses. Chloroform is a carcinogen and may damage the liver and kidneys.

DEPC is a potent protein denaturant and is a suspected carcinogen. It should be handled with care. Wear gloves and work in a chemical fume hood. Point the bottle away from you when opening it; internal pressure can lead to splattering!

Ethidium bromide is a powerful mutagen and moderately toxic. Wear gloves when working with solutions that contain this dye. After use, decontaminate and discard according to local safety office regulations.

Phenol is highly corrosive and can cause severe burns. Wear gloves, protective clothing, and safety glasses and work in a chemical fume hood. Any areas of skin that come in contact with phenol should be rinsed with a large volume of water or PEG 400 and washed with soap and water; do not use ethanol!

METHODS

All of the glassware and the solutions must be RNase-free. Bake the glassware for 5 hours at 250°C. Prepare the reagents using water incubated with 0.1% (v/v) DEPC for 1 hour at 37°C and then autoclave for 1 hour at 121°C. Wear gloves when handling reagents from freshly opened bottles.

1. Grow cells in 1–5 liters of complete medium (see Protocol 6). Harvest the cells during exponential growth by centrifugation in a precooled JA 20 rotor at 4300*g* for 20 minutes at 4°C. The pellet should be 0.5–1.0 g wet weight.

2. Wash rapidly in 2 ml of TE buffer at 4°C.

3. Resuspend the cells in 5 ml of Extraction Buffer and mix thoroughly by inversion. Wait 5–10 minutes at 25°C to achieve complete lysis.

4. Add 0.5 ml of 2 M sodium acetate, 5 ml of TE-saturated phenol, and 1 ml of chloroform/isoamyl alcohol (24:1). Vortex briefly and centrifuge at 12,000*g* for 20 minutes at 4°C.

5. Remove the aqueous phase and extract twice with phenol/chloroform (1:1 v/v) and once with chloroform/isoamyl alcohol (24:1 v/v).

6. Recover the aqueous phase in a 15-ml Corex tube, add 2.5 volumes of 100% ethanol, and 1/10 volume of 3 M sodium acetate. Incubate 30 minutes at –20°C.

7. Centrifuge at 12,000*g* for 30 minutes at 4°C. Wash the pellet three times with 70% ethanol and air-dry.

8. Resuspend the dry pellet in TE buffer to a final volume of 1 ml.

9. Judge the quality of the RNA on a mini-agarose gel (1.2% agarose in TBE, pH 8.0). Discrete bands of 5S, 16S, and 23S rRNAs should be visible with ethidium bromide staining.

COMMENTARY

- Starting with 0.5 g of *Pyrococcus furiosus* cells, we obtain 1.5–1.7 mg of RNA with an A_{260}/A_{280} ratio of 2.00 or above. This procedure has also been performed on *Thermococcus litoralis, Pyrococcus endeavori,* and a new hyperthermophilic isolate, JA1.

- This RNA can be used for Northern blot analysis (DiRuggiero et al. 1993), primer extension, and cDNA cloning. For Northern blot analysis, the RNA is electrophoresed on thin 1.2% agarose/2.2 M formaldehyde surface tension gels (Rosen et al. 1990). Northern blots and dot blots are performed on nylon membranes by capillary transfer (DiRuggiero et al. 1993).

REFERENCES

DiRuggiero, J., L.A. Achenbach, S.H. Brown, R.M. Kelly, and F.T. Robb. 1993. Regulation of ribosomal RNA transcription by growth rate of the hyperthermophilic archaeon, *Pyrococcus furiosus. FEMS Microbiol. Lett.* **111:** 159–164.

Rosen, K.M., E.D. Lamperti, and L. Villa-Komaroff. 1990. Optimizing the Northern blot procedure. *Biotechnology* **8:** 398–403.

Reliable Amplification of Hyperthermophilic Archaeal 16S rRNA Genes by the Polymerase Chain Reaction

A.-L. Reysenbach and N.R. Pace

The polymerase chain reaction (PCR) (Saiki et al. 1988) has been extensively used for isolation of rRNA genes from pure and mixed cultures of bacteria, for identification, and for phylogenetic analysis (Giovannoni 1990; Ward et al. 1992). We have found that amplification of hyperthermophile rRNA genes may sometimes be problematic, with either amplification of nonspecific products or in the case of mixed community DNA, the preferential amplification of some rRNA genes (Reysenbach et al. 1992). These problems may be due to secondary structure in the template, which can limit denaturation of the template and binding of the primer. Mild denaturants such as acetamide have been shown to enhance PCR with some templates (Oleson and You 1989; You 1992). Described here is a reliable procedure, using acetamide as a mild denaturant, for minimizing selective amplification of rRNA genes in hyperthermophilic Archaea.

Numerous descriptions of PCR methods are available (see, e.g., Coen and Scharf 1989; Sambrook et al. 1989). The reaction conditions listed here are based on described protocols (Ponce and Micol 1992; Reysenbach et al. 1992).

MATERIALS

Use sterile, double-distilled H_2O to prepare all reagents.

Microfuge (Eppendorf 5415C)
Microfuge tubes (0.5 ml; Sarstedt 72.699)
Thermal cycler
UV light source (short wavelength or incident)
Taq DNA polymerase (Perkin Elmer N8010060)
Nonidet P-40 (1%, v/v) (Sigma N 6507)
Acetamide (50%, w/v) (Sigma A 0500)
Ethidium bromide
Phenol/chloroform/isomyl alcohol (25:24:1)
Mineral oil
Ethanol (70%)
dNTPs: 1.5 mM (each) dATP, dCTP, dGTP, dTTP in TE buffer (pH 7.0)
Forward and reverse primers at 100 mg/ml each (for selection of primers, see Commentary)
10x Reaction buffer: 300 mM Tricine (pH 8.4), 500 mM KCl, 15 mM $MgCl_2$
TE buffer (pH 8.0): 10 mM Tris-HCl (pH 8.0), 1 mM EDTA (pH 8.0)

Cautions: Chloroform is irritating to the skin, eyes, mucous membranes, and respiratory tract. Work in a chemical fume hood and wear gloves and safety glasses. Chloroform is a carcinogen and may damage the liver and kidneys.

Ethidium bromide is a powerful mutagen and moderately toxic. Wear gloves when working with solutions that contain this dye. After use, decontaminate and discard according to local safety office regulations.

Phenol is highly corrosive and can cause severe burns. Wear gloves, protective clothing, and safety glasses and work in a chemical fume hood. Any areas of skin that come in contact with phenol should be rinsed with a large volume of water or PEG 400 and washed with soap and water; do not use ethanol!

Ultraviolet radiation is dangerous, particularly to the eyes. To minimize exposure, make sure that the ultraviolet light source is adequately shielded and wear protective goggles or a full safety mask that efficiently blocks ultraviolet light.

METHODS

1. Store all sterile reagents at –20°C and defrost just prior to use. Mix well and keep all reagents on ice.

2. Prepare the reaction mixture in the following order in sterile 0.5-ml microfuge tubes. Mix the reaction tube gently but thoroughly between each addition.

Double-distilled sterile H_2O to adjust final volume	
1% Nonidet P-40	5 µl
10x reaction buffer	10 µl
50% acetamide	10 µl
dNTPs	10 µl
Forward primer	2 µl
Reverse primer	2 µl
DNA	10–100 ng
Taq DNA polymerase	1 unit
Total	100 µl

Overlay the reactions with 100 µl of mineral oil. *Note:* A reaction cocktail including all ingredients except *Taq* polymerase can be prepared when a large number of reactions are envisioned. Always prepare a control lacking template DNA, as universally conserved primers also amplify contaminants in the enzyme, buffers, etc. (Schmidt et al. 1991). Contamination can be reduced by using positive-displacement pipettors and by exposing the reaction tubes to ultraviolet light for 15 minutes prior to the addition of DNA and *Taq* polymerase (Kwok and Higuchi 1989).

3. The following is an example of a program to use in a thermal cycler:

94°C for 4 minutes
30 cycles of
    92°C, 1 minute 30 seconds
    50°C, 1 minute 30 seconds
    72°C, 2 minutes, which is extended for 5 seconds after each cycle.

4. Remove 10 μl of the reactions, taking care not to draw up any mineral oil. Resolve products by electrophoresis in a 1% agarose gel, stain with ethidium bromide, and visualize the products using short-wavelength (254 nm) transmitted or incident UV light.

5. Remove the remaining reaction volume, taking care not to draw mineral oil. Extract the amplified DNA with phenol/chloroform/isoamyl alcohol (25:24:1). Add sodium acetate to 0.3 M and precipitate the DNA with 2 volumes of ethanol, wash with 70% ethanol, dry the DNA, and resuspend in TE buffer (pH 8.0). *Note:* To sequence the products directly, purify the DNA further (see, e.g., Sambrook et al. 1989). If the product is to be cloned and then sequenced, the purity of the DNA as described above is sufficient.

COMMENTARY

Reaction Conditions and Thermal Cycling Parameters

• Strategies for optimization of PCR conditions are discussed thoroughly in a number of excellent reviews (see, e.g., Innis and Gelfand 1990; Ehrlich et al. 1991). The reaction mixture suggested in this protocol differs from that of those commonly used in the addition of Nonidet P-40 and acetamide, and the exclusion of a "carrier" protein. The Nonidet P-40 replaces proteins such as gelatin or bovine serum albumin in preventing adsorption of macromolecules to tube surfaces. The proteins are potential sources for contamination and can be inhibitory to the reaction. The acetamide, as other denaturants (Oleson and You 1989; You 1992), likely maximizes template denaturation and primer annealing and thereby reduces nonspecific priming and selective amplification of certain templates in a mixed DNA composition (Reysenbach et al. 1992). Other approaches to maximize template denaturation and primer annealing have been reported, for instance, "hot starts" (Ehrlich et al. 1991) and alkaline denaturation (Cusi et al. 1992).

• The annealing and extension temperatures and times may be varied. Reducing the extension time to 30 seconds is unlikely to affect the results. Due to the relatively high G + C content of most 16S rRNA genes, it is advisable to use higher annealing temperatures. Reduction of the annealing temperature to 37°C may result in amplification of nonspecific products. Increasing the annealing temperatures somewhat beyond 50°C does not appear to impair or improve the amplification under the conditions described in this protocol.

Template Purity

• Certain reagents, such as sodium dodecyl sulfate (SDS) and sodium acetate, routinely used in preparing genomic DNA, may be still present in the DNA preparation and can decrease the efficiency of the PCR. Additional organic extraction of the DNA and precipitation using ethanol and 2.5 M ammonium acetate often alleviates the decreased PCR efficiency.

Primer Selection

• Table 1 provides a selection of forward and reverse primers for amplifying small subunit rRNA genes. Several "universal" primers (applicable with all phylogenetic domains) are included. Both the forward and reverse primers anneal to rRNA genes, whereas only the reverse primers anneal to the rRNA.

Table 1 Some Small Subunit rRNA Gene Primers for PCR and Sequencing

Primer (*E. coli* nucleotide no.)	Sequence (5′→3′)[a]	Archaea	Bacteria	Eukarya
Reverse primers				
(325–309)	TCAGGCTCCCTCTCCGG			+
(357–341)	CTGCTGCCTCCCGTAGT		+	
(536–519)	GWATTACCGCGGCKGCTG		+	+
(704–690)	TCTACGCATTTCACC		+	
(704–690)	TCCAAGAATTTCACC			+
(926–907)	CCGTCAATTCCTTTRAGTTT	+	+	
(976–958)	YCCGGCGTTGAMTCCAATT	+		
(1115–1100)	AGGGTTGCGCTCGTTG	+/−	+	+
(1115–1100)	GGGTCTCGCTCGTTG	+		
(1216–1200)	TCGTAAGGGCCATGATG		+	
(1209–1195)	GGGCATCACAGACCTG			+
(1407–1391)	GACGGGCGGTGTGTRCA	+	+	+
(1510–1492)	GGTTACCTTGTTACGACTT	+	+	+
Forward primers				
(8–27)	AGAGTTTGATCCTGGCTCAG		+	
(2–21)	TTCCGGTTGATCCYGCCGGA	+		+/−
(50–68)	AACACATGCAAGTCGAACG		+	
(333–348)	TCCAGGCCCTACGGG	+		
(342–357)	CTACGGGRSGCAGCAG		+	
(515–533)	GTGCCAGCMGCCGCGGTAA	+	+	+
(906–922)	GAAACTTAAAKGAATTG	+		
(1042–1060)	GAGAGGWGGTGCATGGCC	+		
(1098–1114)	GGCAACGAGCGMGACCC	+	+	

Primers with polylinkers

Bacterial

(8–27)

*Not*I
GCGGATCCGCGGCCGCTGCAGAGTTTGATCCTGGCTCAG
*Bam*HI *Pst*I

(50–68)

*Not*I
GGATCCGCGGCCGCTGCAGAACACATGCAAGTCGAACG
*Bam*HI *Pst*I

Universal

(515–533)

*Xba*I
GCGGATCCTCTAGACTGCAGTGCCAGCAGCCGCGGTAA
*Bam*HI *Pst*I *Sac*II

(1510–1492)

*Not*I
GGCTCGAGCGGCCGCCCGGGTTACCTTGTTACGACTT
*Xho*I *Xma*I

[a]M = A or C; R = A or G; W = A or T; S = C or G; Y = C or T; K = G or T. + indicates effective for most genera in the indicated groups and +/− indicates functions poorly.

The choice of forward and reverse primers will depend on the goals. If most of the 16S rRNA gene from an archaeal isolate is required, then the 2–21 and 1510–1492 primers are a good combination. For molecular phylogenetic analyses of a complex community, universal primers that select DNA templates from all phylogenetic domains, such as 515–533 and 1510–1492, can be used. In this case, the amplified products must be sorted by cloning, so it is advisable to use at least one primer that is flanked by restriction endonuclease sites. Restriction endonucleases may cleave the product internally; however, this may be overcome by using endonucleases that rarely cleave rRNA genes, such as *Not*I. Alternatively, the amplified products can be cloned into a ddT-tailed vector (Holten and Graham 1991). Design of organism-specific primers is facilitated by the collection of aligned small subunit rRNA sequences available from the Ribosomal Database Project (Larsen et al. 1993).

REFERENCES

Coen, D.M. and S.J. Scharf. 1989. Enzymatic amplification of DNA by the polymerase chain reaction: Standard procedures and optimization. In *Current protocols in molecular biology* (ed. E.M. Ausubel et al.), pp. 15.1.17–15.1.4. John Wiley, New York.

Cusi, M.G., L. Cioe, and G. Rovera. 1992. PCR amplification of GC-rich templates containing palindromic sequences using initial alkali denaturation. *BioTechniques* **12:** 502–504.

Ehrlich, H.A., D. Gelfand, and J.J. Sninsky. 1991. Recent advances in the polymerase chain reaction. *Science* **252:** 1645–1651.

Giovannoni, S. 1990. The polymerase chain reaction. In *Nucleic acid techniques in bacterial systematics* (ed. E. Stackebrandt and M. Goodfellow), pp. 177–203. John Wiley, New York.

Holten, T. A. and M.W. Graham. 1991. A simple and efficient method for direct cloning of PCR products using ddT-tailed vectors. *Nucleic Acids Res.* **19:** 756.

Innis, M.A. and D.H. Gelfand. 1990. Optimization of PCRs. In *PCR protocols. A guide to methods and applications* (ed. M.A. Innis et al.), pp. 3–12. Academic Press, San Diego.

Kwok, S. and R. Higuchi. 1989. Avoiding false positives with PCR. *Nature* **339:** 237–238.

Larsen, N., G.J. Olsen, B.L. Maidak, M.J. McCaughey, R. Overbeek, T.J. Macke, T.L. Marsh and C.R. Woese. 1993. The ribosomal database project. *Nucleic Acids Res.* **21:** 3021–3023.

Oleson, A.E. and L. You. 1989. Program Abstract. In *4th San Diego Conference on DNA probes,* Abstr. 25.

Ponce, M.R. and J. Micol. 1992. PCR amplification of long DNA fragments. *Nucleic Acids Res.* **20:** 623.

Reysenbach, A.-L., L.J. Giver, G.S. Wickham, and N.R. Pace. 1992. Differential amplification of rRNA genes by polymerase chain reaction. *Appl. Environ. Microbiol.* **58:** 3417–3418.

Saiki, R.K., D.H. Gelfand, S. Stoffel, S.J. Scharf, R. Higuchi, G.T. Horn, K.B. Mullis, and H.A. Erlich. 1988. Primer directed enzymatic amplification of DNA with thermostable DNA polymerase. *Science* **239:** 487–491.

Sambrook, J., E.F. Fritsch, and T. Maniatis. 1989. *Molecular cloning: A laboratory manual,* 2nd edition. Cold Spring Harbor Laboratory Press, Cold Spring Harbor, New York.

Schmidt, T.M., B. Pace, and N.R. Pace. 1991. Detection of DNA contamination in *Taq* polymerase. *BioTechniques* **11:** 176–177.

Ward, D.M., M.M. Bateson, R. Weller, and A.L. Ruff-Roberts. 1992. Ribosomal RNA analysis of microorganisms as they occur in nature. *Adv. Microb. Ecol.* **12:** 219–286.

You, L. 1992. "Characterization of insertion sequence-like genetic elements from two subspecies of *Clavibactor michiganensis.*" Ph.D. thesis, North Dakota State University, Fargo.

PROTOCOL 17

Archaeal Introns in the rRNA Genes of Hyperthermophiles

J.Z. Dalgaard, J. Lykke-Andersen, and R.A. Garrett

A separate class of introns, probably specific for the archaeal domain, occurs in genes of extreme halophiles, hyperthermophiles, and possibly methanogens. So far, this type of intron has only been detected in stable RNA genes, i.e., in tRNA genes of extreme halophiles (Daniels et al. 1985) and in tRNA and rRNA genes of hyperthermophiles (Kaine et al. 1983; Kjems and Garrett 1985; Kjems et al. 1989a,b; Dalgaard et al. 1992; Burggraf et al. 1993). The gene location sites of the introns are not conserved in either position or sequence. However, at the RNA level, a "bulge-helix-bulge" structure can always form at the intron-exon junctions (Fig. 1) (Thompson and Daniels 1988, 1990; Kjems and Garrett 1991; Dalgaard and Garrett 1992). This structural motif is recognized by a protein enzyme that cleaves within the two "bulges" (Kjems and Garrett 1988; Thompson and Daniels 1988). The exons are subsequently ligated to create the mature RNA molecule, and, in general, the ends of the excised intron are ligated to generate stable circular molecules (Kjems and Garrett 1988; Thompson and Daniels 1988; Dalgaard and Garrett 1992; Lykke-Andersen and Garrett 1994).

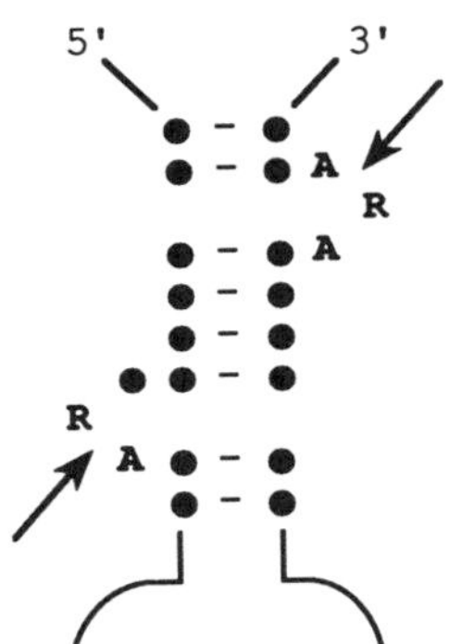

Figure 1 Secondary Structural Motif Conserved at the Intron-Exon Boundaries of Archaeal Introns. The upper two base pairs may not always form in the rRNA introns and in at least one hyperthermophile tRNA intron (Kjems et al. 1989b). Arrows indicate the cleavage sites. Nucleotides in the bulges that are more than 70% conserved are given. (A) Adenine; (R) purine (Kjems and Garrett 1991).

Two-dimensional Polyacrylamide Gel Electrophoresis of Total RNA

The circular RNA introns are present at fairly high levels in the cells, such that they can be stained with silver in two-dimensional polyacrylamide gels of whole-cell extracts (Kjems and Garrett 1988; Dalgaard 1992). The procedure for resolving the circular molecules is based on the principle that the electrophoretic mobility of linear and circular RNA molecules differs at different acrylamide concentrations. When the two dimensions contain different acrylamide concentrations, the linear species run on a diagonal line and the circular molecules migrate above this line (Kjems and Garrett 1988). The method can be used to screen organisms for the presence of stable circular RNA introns (Fig. 2). Total RNA can be prepared from a frozen cell mass using the guanidinium isothiocyanate procedure (Chirgwin et al. 1979), which can be readily scaled down if necessary. After purification, the total RNA is dissolved in double-distilled H_2O, and its concentration is determined (1 A_{260} unit ~35 μg) before storing at –80°C.

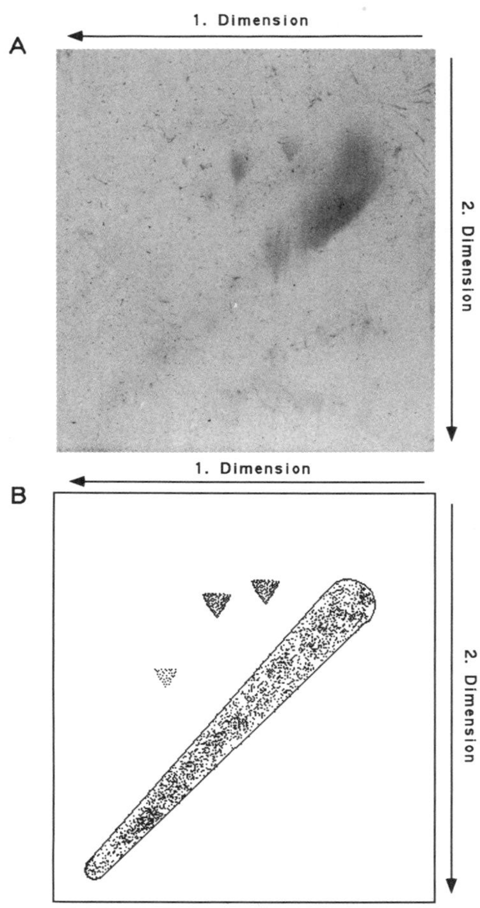

Figure 2 A Two-dimensional Polyacrylamide Gel of Total RNA Extracted from *Pyrobaculum organotrophum* (*A*) Together with an Explanatory Diagram (*B*). The major circular species are indicated in *B*. (Reprinted from Dalgaard 1992.)

MATERIALS

Mini two-dimensional gel apparatus (12 x 12 cm; Hoeffer)
Power source
Syringe (10 ml)
40% acrylamide-bisacrylamide mix
- 96.5 g of acrylamide
- 3.35 g of methylenebisacrylamide
- double-distilled H_2O to 250 ml

10% glycerol, 7 M urea solution
10x TBE
- 108 g of Tris base
- 55 g of boric acid
- 40 ml of 0.5 M EDTA
- double-distilled H_2O to 1 liter

Urea loading buffer
- 0.5x TBE
- 7 M urea
- 0.02% bromophenol blue (Sigma B 5525)
- 0.02% xylene cyanol (Sigma X 4126)

METHODS

1. For casting the gels (12 x 12 cm), pour the first dimension containing 3.5% acrylamide, 7 M urea, and 0.5x TBE. Cast the second dimension as a composite gel. Overlay the lower two thirds of the gel containing 5% acrylamide, 7 M urea, and 1x TBE with water while the gel is polymerizing. Pour the upper third containing 3.5% acrylamide, 7 M urea, and 0.5x TBE.

2. Apply 5–10 μg of total RNA on the gel in a total volume of 5–10 μl of urea loading buffer. Run the first dimension at 150 V for 1.5 hours at 4°C using 0.5x TBE as running buffer.

3. Carefully remove the gel from the glass tube using a 10-ml syringe filled with 7 M urea. Quickly wash the gel in 7 M urea and place on the second gel.

4. Run the second dimension at 90 V for 20 hours at 4°C with 0.5x TBE in the upper chamber and 1x TBE in the lower chamber.

5. Stain the gel with silver (Beidler et al. 1982).

Detection of In Vivo Splicing

During splicing, an intron sequence is removed from a precursor RNA molecule and the exons are ligated. If it is clear from Southern analysis (Sambrook et al. 1989) that the gene is present in only one copy per genome or that only putative intron-containing alleles are present, then the occurrence of a splicing reaction can be established by sequencing the RNA species present in the cell extracts using reverse transcriptase and chain termination (Sanger et al. 1977). The circularization of an intron can be monitored in the same way as exon ligation (Fig. 3).

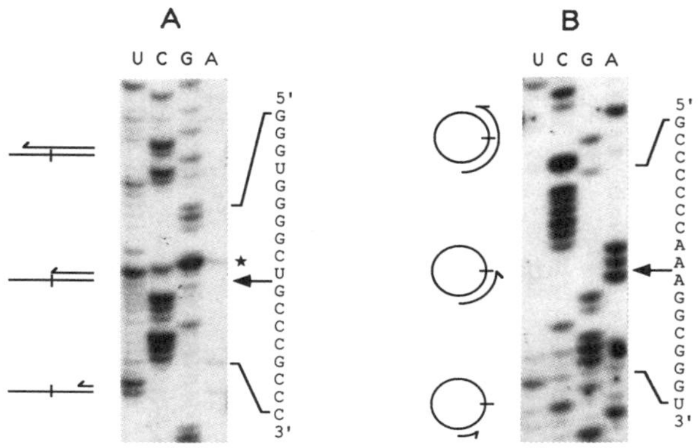

Figure 3 Sequencing Gel of the Mature RNA Transcript (*A*) and the Circularized RNA Intron Molecule (*B*). Both RNAs derive from the pre-23S rRNA gene of *Pyrobaculum organotrophum* (Dalgaard and Garrett 1992). The asterisk indicates sequence compression in *A*.

MATERIALS

Microcentrifuge
Sequencing gel electrophoresis apparatus
Power source
X-ray film (Fugi, Rx 18 x 43 cm)
Water bath (50°C)
Total RNA
Deoxyoligonucleotide primer
[γ-^{32}P]ATP T4 polynucleotide kinase (Amersham E2021Y)
AMV reverse transcriptase (Life Sciences AMV007)
2.5 mM dNTPs (pH 7.0)
0.25 mM ddATP
0.25 mM ddGTP
0.20 mM ddCTP
0.20 mM ddTTP
Sodium acetate (0.3 M, pH 6.0)
Ethanol (absolute and 80%)
Formamide loading buffer
- 80% deionized formamide
- 10 mM NaOH
- 1 mM EDTA
- 0.02% bromophenol blue (Sigma B 5525)
- 0.02% xylene cyanol (Sigma X 4126

25x REV buffer
- 1.25 M Tris-HCl (pH 8.4)
- 250 mM $MgCl_2$
- 50 mM DTT (dithiothreitol)

10x TKE buffer
- 100 mM Tris-HCl (pH 6.9)
- 400 mM KCl
- 5 mM EDTA

METHODS

1. Synthesize an oligodeoxynucleotide primer (17–22 mer) that can anneal to the mature RNA at least 50 nucleotides 3′ to the putative intron insertion site. End-label the primer using [γ-^{32}P]ATP and T4 polynucleotide kinase (Sambrook et al. 1989).

2. Anneal 2 pmoles of end-labeled primer to 5 μg of total RNA in a total volume of 24 μl of 1x TKE buffer. Denature the sample for 30 seconds at 95°C and transfer to a 50°C water bath. Incubate for 20 minutes, spin for 30 seconds, and split the sample into four aliquots. Label the four samples A, U, G, and C and store the tubes on ice.

3. Add to each of the four tubes: 0.8 μl of one of the ddNTP dilutions and 4 μl of extension mixture (2.8 μl of double-distilled H_2O, 0.8 μl of 2.5 mM dNTPs, 0.4 μl of 25x REV buffer, and 1 unit of AMV reverse transcriptase). Incubate for 30 minutes at 37°C.

4. Add 40 μl of 0.3 M sodium acetate (pH 6.0) and 130 μl of absolute ethanol. Place on dry ice for 30 minutes. Centrifuge at 14,000 rpm for 20 minutes, remove ethanol, and wash with 80% ethanol (–20°C). Dry the pellet and dissolve in formamide loading buffer at a radioactivity concentration of approximately 100 cps/μl. Denature the sample for 2 minutes at 95°C and load 1 μl on a 6% polyacryl-amide sequencing gel. After electrophoresis, autoradiograph the gel.

In Vitro Splicing of the Transcript from Cloned Intron-containing Genes

To monitor intron splicing in vitro (Kjems and Garrett 1988, 1991), the intron and flanking exons are cloned into a bacterial plasmid vector (e.g., pBluescript) downstream from a T7 promoter (Studier and Moffatt 1986). A T7 RNA polymerase transcript is then generated that can be cleaved specifically at the exon-intron junctions by crude cell extracts. Comparison of the nucleotide sequence before and after splicing will reveal the exact insertion site.

Preparation of Crude Cell Extracts

MATERIALS

Cells (see Protocols 1–4 for culture methods)
Glass beads (sigma G 8772)
Vortex
Microcentrifuge
Sorvall SS-34 rotor (DuPont Medical Products)
Sephadex G-50 column (Pharmacia 17-0042-01)
10% DMSO (dimethylsulfoxide)
0.1% Triton X-100
PMSF (phenylmethylsulfonyl fluoride) (60 μg/ml)
DNA template
7.5 mM rNTPs

T7 RNA polymerase (Promega P2075)
Crude cell extract
Carrier RNA (e.g., yeast tRNA)
T7 polymerase transcript
Sodium acetate (0.3 M, pH 6.0)
Phenol (equilibrated with TE)
Chloroform
Ethanol (absolute and 80% v/v)
Lysis buffer
- 100 mM Tris-HCl (pH 7.5)
- 50 mM NaCl
- 10 mM $MgCl_2$
- 0.2 mM EDTA
- 2 mM DTT (dithiothreitol)
- 10% glycerol

10x Splice buffer
- 500 mM Tris-HCl (pH 7.5)
- 100 mM $MgCl_2$ (for *Dc. mobilis*, use 50 mM EDTA instead of $MgCl_2$)
- 1 M NaCl

10x T7 RNA polymerase buffer
- 400 mM Tris-HCl (pH 8.0)
- 60 mM $MgCl_2$
- 50 mM DTT
- 10 mM spermidine

TM buffer
- 10 mM Tris-HCl (pH 7.5)
- 1 mM $MgCl_2$

Cautions: Chloroform is irritating to the skin, eyes, mucous membranes, and respiratory tract. Work in a chemical fume hood and wear gloves and safety glasses. Chloroform is a carcinogen and may damage the liver and kidneys.

DMSO is harmful if inhaled or absorbed through the skin. Wear gloves, safety glasses, and a mask and work in a chemical fume hood. DMSO is also combustible. Store in a tightly closed container.

Phenol is highly corrosive and can cause severe burns. Wear gloves, protective clothing, and safety glasses and work in a chemical fume hood. Any areas of skin that come in contact with phenol should be rinsed with a large volume of water or PEG 400 and washed with soap and water; do not use ethanol!

PMSF is extremely destructive to the mucous membranes of the respiratory tract, eyes, and skin. It may be fatal if inhaled, swallowed, or absorbed through the skin. It is a highly toxic cholinesterase inhibitor. Wear gloves and safety glasses and work in a chemical fume hood.

METHODS

Preparation of Crude Extracts

1. Suspend 1 g of cells in 2 ml of lysis buffer and add 2 ml of glass beads. Vortex for 15 minutes at 4°C. Centrifuge at 10,000 rpm for 15 minutes in a Sorvall SS-34 rotor.

2. Remove the supernatant and centrifuge at 18,000 rpm for 45 minutes. Store the supernatant in aliquots at –80°C.

T7 RNA Polymerase Transcription

1. Linearize the plasmid by cutting downstream from the intron at a suitable restriction site.

2. In a total volume of 20 μl, mix 0.5 pmole of DNA template (linearized), 2 μl of 10x T7 polymerase buffer, and 2 μl of 7.5 mM rNTPs (if randomly labeled RNA is needed, add some [α-^{32}P]rUTP). Add 10 units of T7 RNA polymerase and incubate for 90 minutes at 37°C. This should generate 5–10 μg of RNA.

3. Separate the RNA from unpolymerized nucleotides by spinning through a 1-ml Sephadex G-50 column (Sambrook et al. 1989).

In Vitro Splicing

1. Include in a total volume of 20 μl of mix 1 μl of crude cell extract, 20 μg of carrier RNA, 5–10 pmoles of T7 polymerase transcript, and 2 μl of 10x Splice buffer. Incubate for 15 minutes at 56°C.

2. Add 130 μl of 0.3 M sodium acetate (pH 6.0). Extract twice with phenol, once with phenol/chloroform, and once with chloroform.

3. Add 2.5 volumes absolute ethanol (–20°C) and place on dry ice for 30 minutes. Centrifuge at 14,000 rpm for 20 minutes, remove ethanol, and wash with 80% ethanol (–20°C). Dry pellet and dissolve in TM buffer.

Identifying RNA Cleavage Sites

1. Monitor the cleavage sites by using the primer extension method described above (Detection of In Vivo Splicing, p. 109), making use of the inefficient ligation that occurs in vitro.

2. Hybridize the oligodeoxynucleotide primers to the RNA at least 50 nucleotides downstream from each putative cleavage site (Fig. 4).

3. Carry out primer extension as sequencing reactions (i) on the unspliced transcript and—omitting the ddNTP—(ii) on the spliced transcript, and (iii) on the unspliced transcript. Run the samples on a sequencing gel. If the intron is short (<150 nucleotides), then both sites can be monitored using a single primer by lowering the amount of cell extract employed in the splicing reaction that leads to partial cleavage at each site.

Figure 4 Diagram Showing the Approximate Locations of Oligonucleotide Primers in Relation to the Intron Cleavage Sites.

Proteins Encoded by rRNA Introns from Hyperthermophiles

A few introns from hyperthermophiles have been characterized that contain open reading frames (ORFs) (Kjems and Garrett 1985; Dalgaard and Garrett 1992; Burggraf et al. 1993). Two of these proteins have been expressed in vitro and shown to possess site-specific DNA endonuclease activity (Dalgaard et al. 1993; Lykke-Andersen et al. 1994). The homing endonucleases recognize and cleave the intron-minus version of the gene at the position where the intron is inserted. This type of site-specific endonuclease is also encoded by group I introns from eukaryotes and bacteriophages and can be formed by protein introns in Archaea, bacteria, and eukaryotes (Lambowitz and Belfort 1993). The site-specific endonuclease activity is difficult to detect in crude cell extracts because unspecific degradation takes place. Therefore, it is advantageous to clone the ORF into a vector downstream from a suitable promoter such that it can be transcribed and translated in vitro.

MATERIALS

Vertical electrophoresis apparatus (homemade 20 x 20 cm)
Power source
Vacuum gel drier (Hoeffer SE 1160)
Horizontal electrophoresis apparatus
X-ray film (Fugi Rx 13 x 18 cm)
Microcentrifuge
Sequencing gel electrophoresis apparatus (20 x 80 cm Biolabs)
Rabbit reticulocyte lysate (Promega L4960)
Amino acid mixture (1 mM, minus methionine) (Promega L9961)
[^{35}S]methionine (1200 Ci/mmole at 10 mCi/ml)
12% polyacrylamide SDS gel
Agarose
Acetic acid (10%)
Ethanol (10%, 80%, 100%)
Chloroform
Sodium acetate (3 M, pH 6.0)
Phenol (equilibrated with TE buffer)
10 pmoles/μl deoxyoligonucleotide primer
2.5 mM dNTPs
[α-^{35}S]dATP (1000 Ci/mmole at 10 mCi/ml; Amersham 531304)
10x Assay buffer
- 500 mM Tris-acetate (pH 8.0)
- 100 mM magnesium acetate
- 100 mM ammonium acetate
- 10 mM DTT (dithiothreitol)

Sequenase sequencing kit (USB 70770) containing
- ddATP mix
- ddTTP mix
- ddGTP mix
- ddCTP mix
- 0.1 M DTT
- Sequenase

- Labeling mix
- Formamide loading buffer
  - 80% deionized formamide
  - 10 mM NaOH
  - 1 mM EDTA
  - 0.02% bromophenol blue (Sigma B 5525)
  - 0.02% xylene cyanol (Sigma X 4126)
- 5x Sequenase buffer
- Sequenase dilution buffer

SDS sample buffer
- 100 mM Tris-HCl (pH 6.8)
- 200 mM DTT (dithiothreitol)
- 4% SDS (sodium dodecyl sulfate)
- 0.2% bromophenol blue
- 20% glycerol

Stop buffer
- 75 mM EDTA (pH 8.0)
- 0.02% bromophenol blue
- 0.02% xylene cyanol
- 0.5% SDS
- 10% Ficoll (Sigma F 2637)

TE buffer
- 10 mM Tris-HCl (pH 8.0)
- 0.1 mM EDTA

Caution: Chloroform is irritating to the skin, eyes, mucous membranes, and respiratory tract. Work in a chemical fume hood and wear gloves and safety glasses. Chloroform is a carcinogen and may damage the liver and kidneys.

Phenol is highly corrosive and can cause severe burns. Wear gloves, protective clothing, and safety glasses and work in a chemical fume hood. Any areas of skin that come in contact with phenol should be rinsed with a large volume of water or PEG 400 and washed with soap and water; do not use ethanol!

METHODS

In Vitro Translation

1. Carry out in vitro transcription with T7 RNA polymerase as described above (see p. 113).

2. For in vitro translation with rabbit reticulocyte lysate, mix 35 μl of lysate, 7 μl of double-distilled H_2O, 1 μl of 1 mM amino acid mixture (minus methionine), 2 μg of RNA substrate, and 5 μl of [^{35}S]methionine on ice. Include a control sample with double-distilled H_2O added instead of the RNA substrate. Incubate for 60 minutes at 30°C. Add 15 μl of glycerol and store at –20°C.

3. Analyze the translation product(s) by removing a 10-μl aliquot and mixing it with 40 μl of SDS sample buffer. Heat the sample for 2 minutes at 90°C and load 10 μl on a 12% polyacrylamide SDS gel. After electrophoresis, dry the gel and autoradiograph.

Detection of DNA Cleavage Activity

Cleavage activity can be detected using the protein synthesized in vitro in the rabbit reticulocyte lysate. To make the assay as sensitive as possible, end-label the DNA substrate using one of various methods (Sambrook et al. 1989).

1. Mix 100 cpm of labeled DNA (~10 ng) with 5 µl of 10x Assay buffer and double-distilled H_2O to 50 µl. Add 1 µl of in vitro translation mix (see preceding section) and include a control with the unprogrammed in vitro translation mix. Incubate for 1 hour at 75°C. Add 10 µl of Stop buffer.

2. Load an aliquot on a 1% agarose gel. After electrophoresis, fix the gel in 10% acetic acid and 10% ethanol for 20 minutes. Dry the gel (room temperature) and autoradiograph.

Cleavage Site Mapping

This method, developed in Dr. M. Belfort's laboratory (Chirgwin et al. 1979; Wenzlau et al. 1989), has been successfully used to map the cleavage site of the endonuclease encoded by the rRNA intron of *Dc. mobilis* (Dalgaard et al. 1993). Synthesize the appropriate deoxyoligonucleotide primers for sequencing both DNA strands across the cleavage site.

1. Mix 2 µg of double-stranded DNA prepared for sequencing with 2 µl of 10 pmoles/µl oligonucleotide and 3 µl of 5x Sequenase buffer. Adjust volume to 20 µl with double-distilled H_2O. Incubate for 30 minutes at 37°C.

2. Simultaneously add 2.5 µl of 2.5 mM dNTPs to two tubes labeled 1 and 2. Prepare four tubes with 2.5 µl of ddATP mix, ddCTP mix, ddGTP mix, and ddTTP mix, respectively, for the ordinary sequencing reactions (a total of six tubes).

3. Dilute 1.5 µl of Sequenase with Sequenase dilution buffer (1:5) and maintain at 0°C. Add the following mix to the annealed oligonucleotide-DNA sample: 5 µl of 0.1 M DTT, 10 µl of 1:5 of diluted labeling mix, 2.5 µl of [^{35}S]dATP, and 10 µl of diluted Sequenase. Incubate for 5 minutes at 37°C.

4. Prewarm the six tubes and add 3.5 µl of the sample to each of the six tubes. Incubate for 5 minutes at 37°C.

5. Add 3.5 µl of formamide loading buffer to the four sequencing reactions and store on ice. To tubes 1 and 2, add 3.0 µl of 10x Assay buffer, 16 µl of double-distilled H_2O, and 1 µl of programmed and unprogrammed translation mixes, respectively. Incubate for 20 minutes at 75°C.

6. Extract samples 1 and 2 with phenol/chloroform. Add 3 µl of 3 M sodium acetate (pH 6.0) and 60 µl of 100% ethanol. Precipitate for 30 minutes at –80°C. Spin pellets at 15,000 rpm for 15 minutes at 4°C. Wash the pellets in 80% ethanol. Dry in a vacuum centrifuge and redissolve the pellets in 25 µl of formamide loading buffer.

7. Denature all the samples for 2 minutes at 95°C and load 2.5 µl of the sequencing reactions and 0.5–5 µl of samples 1 and 2 on a sequencing gel.

ACKNOWLEDGMENT The research performed in our laboratory and J.Z.D. were supported by grants from the Danish Natural Science Research Council. J.L.-A. received a grant from Copenhagen University.

REFERENCES

Beidler, J.L., P.R. Hilliard, and R.L. Rill. 1982. Ultrasensitive staining of nucleic acids with silver. *Anal. Biochem.* **126:** 374–380.

Bell-Pedersen, D., S. Quirk, J. Clyman, and M. Belfort. 1990. Intron mobility in phage T4 is dependent upon a distinctive class of endonucleases and independent of the DNA sequence encoding the intron core: Mechanistic and evolutionary implications. *Nucleic Acids Res.* **18:** 3763–3770.

Burggraf, S., N. Larsen, C.R. Woese, and K.O. Stetter. 1993. An intron within the 16S rRNA gene of the archaeon *Pyrobaculum aerophilum. Proc. Natl. Acad. Sci.* **90:** 2547–2550.

Chirgwin, J.M., A.E. Przybyla, R.J. MacDonald, and W.J. Rutter. 1979. Isolation of biologically active ribonucleic acid from sources enriched in ribonucleases. *Biochemistry* **18:** 5294–5299.

Dalgaard, J.Z. 1992. "Archaeal introns." Speciale thesis, Copenhagen University, Denmark.

Dalgaard, J.Z. and R.A. Garrett. 1992. Protein-coding introns from the 23S rRNA-encoding gene form stable circles in the hyperthermophilic archaeon *Pyrobaculum organotrophum. Gene* **121:** 103–110.

Dalgaard, J.Z., R.A. Garrett, and M. Belfort. 1993. A site-specific DNase encoded by a typical archaeal intron. *Proc. Natl. Acad. Sci.* **90:** 5414–5417.

Daniels, C.J., R. Gupta, and W.F. Doolittle. 1985. Transcription of a large intron in the $tRNA^{Trp}$ gene of an archaebacterium *Halobacterium volcanii. J. Biol. Chem.* **260:** 3132–3137.

Kaine, B.P., R. Gupta, and C.R. Woese. 1983. Putative introns in tRNA genes of prokaryotes. *Proc. Natl. Acad. Sci.* **80:** 3309–3312.

Kjems, J. and R.A. Garrett. 1985. An intron in the 23S ribosomal RNA gene of the archaebacterium *Desulfurococcus mobilis. Nature* **318:** 675–677.

———. 1988. Novel splicing mechanism for the ribosomal RNA intron in the archaebacterium *Desulfurococcus mobilis. Cell* **54:** 693–703.

———. 1991. rRNA introns in archaea and evidence for RNA conformational changes associated with splicing. *Proc. Natl. Acad. Sci.* **88:** 439–443.

Kjems, J., J. Jensen, T. Olesen, and R.A. Garrett. 1989a. Comparison of transfer RNA and ribosomal RNA intron splicing in the extreme thermophile and archaebacterium *Desulfurococcus mobilis. Can. J. Microbiol.* **35:** 210–214.

Kjems, J., H. Leffers, T. Olesen, and R.A. Garrett. 1989b. A unique tRNA intron in the variable loop of the extreme thermophile *Thermofilum pendens* and its possible evolutionary implications. *J. Biol. Chem.* **264:** 17834–17837.

Lambowitz, A.M. and M. Belfort. 1993. Introns as mobile genetic elements. *Annu. Rev. Biochem.* **62:** 587–622.

Lykke-Andersen, J. and R.A. Garrett 1994. Structural characteristics of the stable RNA introns of archaeal hyperthermophiles and their splicing junctions. *J. Mol. Biol.* **243:** 846–853.

Lykke-Andersen, J., H.P. Thi-Ngoc, and R.A. Garrett. 1994. DNA substrate specificity and cleavage kinetics of an archaeal homing type endonuclease from *Pyrobaculum organotrophum. Nucleic Acids Res.* **22:** 4583-4590.

Sambrook J., E.F. Fritsch, and T. Maniatis. 1989. *Molecular cloning: A laboratory manual,* 2nd edition. Cold Spring Harbor Laboratory Press, Cold Spring Harbor, New York.

Sanger, F., S. Nicklen, and A.R. Coulson. 1977. DNA sequencing with chain terminating inhibitors. *Proc. Natl. Acad. Sci.* **74:** 5463–5467.

Studier, F.W. and B.A. Moffatt. 1986. Use of bacteriophage T7 RNA polymerase to direct selective high-level expression of cloned genes. *J. Mol. Biol.* **189:** 113–130.

Thompson, L.D. and C.J. Daniels. 1988. A $tRNA^{Trp}$ intron endonuclease from *Halobacterium volcanii.* Unique substrate recognition properties. *J. Biol. Chem.* **263:** 17951–17959.

———. 1990. Recognition of exon-intron boundaries by the *Halobacterium volcanii* tRNA intron endonuclease. *J. Biol. Chem.* **265:** 18104–18111.

Thompson, L.D., L.D. Brandon, D.T. Niewlandt, and C.J. Daniels. 1989. tRNA intron processing in the halophilic archaebacteria. *J. Biochem.* **35:** 36–42.

Wenzlau, J.M., R.J. Saldanha, R.A. Butow, and P.S. Perlman. 1989. A latent intron-encoded maturase is also an endonuclease needed for intron mobility. *Cell* **56:** 421–430.

PROTOCOL 18

Generation of Subtraction Probes for Isolation of Specific Genes in Thermophilic Archaea

K.A. Robinson, F.T. Robb, and H.J. Schreier

The creation of a subtractive probe is useful for isolating genes with related functions from a genomic or cDNA library. Currently, hyperthermophilic Archaea cannot be mutagenized due to problems of selection. This severely limits the study of functional genes. This technique is extremely useful in enriching for low-abundance, rare, or induced mRNAs. The majority of RNA extracted from Archaea is ribosomal RNA; therefore, the removal of ribosomal sequences limits the number of clones that need to be screened in a library.

MATERIALS

All reagents must be RNase-free. It is essential that all procedures be carried out under RNase-free conditions.

Sunlamp (GE RSM 275 W or equivalent)
Microcentrifuge tubes with screw caps (1.5 ml; VWR 20170-214)
Vacuum centrifuge (Savant)
Photobiotin acetate (Vector Labs SP-1000)
Tris (0.1 M, pH 9.0)
2-butanol (water-saturated)
NaOAc (2 M)
Ethanol (70% and 100%)
[α-^{32}P]dATP (10 mCi/ml, 3000 Ci/mmole)
MMLV reverse transcriptase (250 units/μl; USB 70456)
Streptavidin (1 mg/ml in distilled H_2O)
β-mercaptoethanol (0.7 M; Sigma M 6250)
MeHgOH (100 mM)
Random primer (1 mg/ml; Promega C1181)
Placental RNase inhibitor (20 units/μl; Promega N2111)
100 mM stocks of dNTPs
NaOH (0.2 N)
TCA (trichloroacetic acid)
Phenol/chloroform
DEPC (diethyl pyrocarbonate)-treated H_2O
HEPES/EDTA buffer: 10 mM HEPES, 10 mM EDTA (pH 7.4)

2x Hybridization buffer

PIPES (pH 6.4)	40 mM
EDTA pH 8	1 mM
NaCl	400 mM
80% formamide	

5x RT buffer

Tris pH 7.6	250 mM
KCl	375 mM
$MgCl_2$	15 mM
DTT (dithiothreitol)	50 mM

Cautions: β-Mercaptoethanol may be fatal if inhaled or absorbed through the skin and is harmful if swallowed. High concentrations are extremely destructive to the mucous membranes, upper respiratory tract, skin, and eyes. Wear gloves and safety glasses and work in a chemical fume hood.

Chloroform is irritating to the skin, eyes, mucous membranes, and respiratory tract. Work in a chemical fume hood and wear gloves and safety glasses. Chloroform is a carcinogen and may damage the liver and kidneys.

DEPC is a potent protein denaturant and is suspected to be a carcinogen. It should be handled with care. Wear gloves and work in a chemical fume hood. Point the bottle away from you when opening it; internal pressure can lead to splattering!

Phenol is highly corrosive and can cause severe burns. Wear gloves, protective clothing, and safety glasses and work in a chemical fume hood. Rinse any areas of skin that come in contact with phenol with a large volume of water or PEG 400 and wash with soap and water; do not use ethanol!

TCA is highly caustic. Always wear gloves when handling this compound.

METHODS

Removal of Ribosomal Genes from Total RNA

1. Extract total RNA from cells grown under both inducing and noninducing conditions following the methods outlined in Protocol 15.

2. Carry out the following procedure in a darkroom, under minimal light. In a screw-cap microcentrifuge tube, combine 80 μg of denatured rRNA-coding sequences contained in plasmid DNA with a freshly prepared 1 mg/ml solution of photobiotin acetate. Place this tube, with its cap on, in an ice slurry approximately 10 cm below a 275-W sunlamp. Irradiate the contents of the tube for 3 minutes. If a sunlamp is not available, use a 300-W reflector lamp for irradiation for 30 minutes on ice.

3. Add 0.1 M Tris (pH 9.0) to a final volume of 200 μl.

4. Extract this solution with 200 μl of water-saturated 2-butanol. Remove the upper 2-butanol layer and discard. Repeat 2-butanol extractions until the lower aqueous layer appears clear.

5. Precipitate the photobiotinylated DNA with 0.1 volume of 2 M NaOAc and 2.5 volumes of 100% ethanol. Wash once with 70% ethanol, dry the pellet, and resuspend in 30 μl of DEPC-treated H_2O.

6. Add 30 μl of photobiotinylated DNA to 10 μg of total RNA extracted from an induced culture. Coprecipitate with 0.1 volume of 2 M NaOAc and 2.5 volumes of 100% ethanol. Wash once with 70% ethanol and dry pellet in a vacuum centrifuge.

7. Resuspend the pellet in 10 μl of DEPC-treated H_2O and add 10 μl of 2x Hybridization buffer. Heat for 1 minute at 100°C and allow to hybridize overnight in a water bath at 68°C.

8. Remove hybridization from water bath and incubate for 5 minutes at 55°C and then on ice.

9. Add 10 ml of streptavidin (1 mg/ml). Incubate for 10 minutes at room temperature.

10. Add 60 μl of phenol/chloroform, vortex, spin in a microcentrifuge for 2 minutes, and transfer the upper aqueous phase to a new tube. The streptavidin will bring down biotinylated nucleic acids into the interphase, so take care to avoid the interphase.

11. Back-extract the organic layer by adding 50 μl of 10 mM HEPES/EDTA (pH 7.4) buffer. Combine the aqueous layer with the layer collected in the previous step.

12. Add 10 μl of streptavidin to the combined aqueous layers and incubate for 5 minutes at room temperature.

13. Add 120 μl of phenol/chloroform, vortex, spin for 2 minutes in a microcentrifuge, and transfer the upper aqueous layer to a fresh tube.

14. Back-extract the organic layer with 50 μl of 10 mM HEPES/EDTA (pH 7.4). Spin for 2 minutes, remove the upper aqueous layer, and combine it with the aqueous layer from the previous step.

15. Precipitate the aqueous layer with 2 M NaOAc and 100% ethanol. Wash with 70% ethanol, dry the pellet, and resuspend in 10 μl of DEPC-treated H_2O.

16. After removing the majority of ribosomal sequences from total RNA, separate a small aliquot (2 μl) by electrophoresis on a 1.2% agarose gel under RNase-free conditions to check the integrity of the messages left and note the absence of the characteristic 23S, 16S, and 5S bands. This RNA should now be devoid of the majority of rRNA sequences.

Creation of cDNA from a Population of Differentially Expressed Genes

rRNA-free total RNA (5 μg) isolated in the steps above is used to create a $[^{32}P]$dATP-labeled cDNA. First-strand synthesis is performed as described below.

1. Add 5 μg of RNA in a volume of 8 μl to 2 μl of 100 mM MeHgOH. Mix tube lightly (do not vortex) and let stand for 5 minutes at room temperature.

2. Add 2.5 μl of 700 mM β-mercaptoethanol, mix, and place the tube on ice.

3. Add 1 μl of random primer (1 μg/μl), incubate in a water bath for 2 minutes at 65°C, and then chill on ice.

4. Add the following in order:

   1 μl of placental RNase inhibitor (25 units/μl)
   4 μl of 5x RT buffer
   0.2 μl each of 100 mM dGTP, dCTP, dTTP
   0.4 μl of 5 mM dATP
   5 μl of [α-^{32}P]dATP
   2 μl of MMLV reverse transcriptase (250 units/ml)

5. Incubate for 1 hour at 37°C.

6. After the first incubation, remove 0.5 μl of the reaction mix. This will be used to measure the incorporation of the probe, via TCA precipitation.

7. Denature the RNA/cDNA hybrids by heating for 3 minutes at 95°C. Cool on ice and then add another 0.5 μl of MMLV reverse transcriptase. Incubate for 1 hour at 37°C.

8. After the second incubation, remove 0.5 μl of the reaction mix to use for measuring incorporation of the probe.

9. After the second round of synthesis has taken place, add 40 μl of 0.2 N NaOH to the tube. Incubate for 15 minutes at 70°C to hydrolyze the RNA.

10. Precipitate nucleic acids with 7.5 M NH_4OAc and 100% ethanol. Resuspend the pellet in a small volume (20 μl) of DEPC-treated H_2O.

11. If desired, electrophorese a small aliquot of the cDNA on a 1% agarose gel, dry down the gel, and expose it to X-ray film. This can be done to assure complete synthesis has taken place. Once cDNA synthesis is complete, subtraction of undesired transcripts is carried out as described below (steps 1–5).

Hybridization of a Differentially Expressed, Induced cDNA Population to Uninduced, Biotinylated RNA

1. As described in steps 2–5 (p. 120), 1 mg of total RNA extracted from a population of cells not subjected to inducing conditions is photobiotinylated.

2. Coprecipitate biotinylated RNA with the cDNA generated from the induced RNA.

3. Hybridize the RNA/cDNA population overnight at 68°C.

4. Remove RNA/cDNA duplexes via phenol/chloroform extractions, using streptavidin, as described in steps 8–14 (p. 121).

5. Precipitate the remaining sequences with 2 M NaOAc and 100% ethanol. Wash with 70% ethanol, dry the pellet, and resuspend in 50 µl of H_2O. The remaining single-stranded cDNAs are now enriched for low-abundance, rare, or induced messages and can be used directly as a probe to screen genomic or cDNA libraries. Alternatively, clone the cDNAs into suitable vectors for direct DNA sequence determination.

COMMENTARY

- The lack of poly(A) tails in prokaryotes creates difficulty in isolating mRNA away from total RNA for cDNA creation. This method removes the majority of ribosomal genes from the pool of total RNA, ensuring that mainly mRNA will be used for cDNA synthesis. The availability of plasmid or PCR product containing ribosomal coding sequences is necessary for this procedure.

- One of the problems encountered when creating the cDNAs from induced RNA is low incorporation of [^{32}P]dATP. By adjusting the amount of "cold" dATP, it is possible to create a probe with higher specific activity.

- It is important to use a tenfold excess of uninduced RNA over induced cDNAs to ensure that all possible duplexes are formed.

ACKNOWLEDGMENT

This work was supported by the National Science Foundation, DOE, and ONR.

PROTOCOL 19

Isolation of *Sulfolobus acidocaldarius* Mutants

D.W. Grogan

The relatively vigorous aerobic, heterotrophic growth of certain *Sulfolobus* strains (Grogan 1989) enables the adaptation of classical microbial genetic techniques to *Sulfolobus* spp., including methods of mutant isolation. The following procedures are based on well-established techniques. Workers unfamiliar with these techniques should consult a text on microbial genetics (see, e.g., Miller 1972; Davis et al. 1980).

Cultivation of *Sulfolobus acidocaldarius*

MATERIALS

2x Basal Medium for *Sulfolobus acidocaldarius* (see Appendix 2, Medium 9)
Incubator (75°C)
Erlenmeyer flask (1 liter; Fisher 10-040K)
Petri dishes (Fisher 08-757-1000)
Culture vessels (Fisher 03-325-1C or 02-945-10)
Spectrophotometer or colorimeter (e.g., Scientific Products C6800)
Gelrite gellan gum (Kelco division of Merck, distributed by SchweizerHall 46-3070-00)
L-glutamine or tryptone
D-xylose
Gel-solidifying solutions (in distilled H_2O): 2 M Mg_2SO_4, 0.5 M $CaCl_2$
Sulfuric acid (H_2SO_4)

Cautions: 5-fluoro-pyrimidine compounds are toxic. Avoid direct contact. Wear gloves and safety glasses.
Sulfuric acid is extremely corrosive. Wear gloves and safety glasses.

METHODS

1. Make up the trace mineral solution and 2x Basal Medium (see Appendix 2).

2. Weigh 2 g of carbon (D-xylose) and 1 g of nitrogen source (L-glutamine or tryptone) into a 1-liter Erlenmeyer flask and dissolve in 500 ml of 2x Basal Medium.

3. For liquid medium, adjust the pH to 3.5 using H_2SO_4. Dilute with an equal volume of distilled H_2O and inoculate (skip to step 9).

4. For pouring plates, adjust the pH to 2.9 using H_2SO_4 and add 5 ml each of the $MgSO_4$ and $CaCl_2$ solutions; add selective agent, if desired, and warm the solution to approximately 60°C.

5. Weigh 7 g of Gelrite into a 1-liter Erlenmeyer flask. Add 490 ml of distilled H_2O and slowly suspend so as to prevent formation of large lumps.

6. Heat with stirring until solution boils and solid dissolves; turn off heat.

7. As the hot Gelrite solution stirs, slowly add warm, concentrated medium.

8. Pour the hot mixture into petri dishes and leave the covered dishes on the bench overnight to dissipate excess moisture. The yield should be approximately 30 10-cm plates.

9. Incubate cultures at 75–80°C. Seal liquid cultures but leave a large headspace of air; place plates inside humidified containers or plastic bags.

COMMENTARY

• The above simplified medium (see Appendix 2) has been formulated for *S. acidocaldarius* strain DG6 (American Type Culture Collection number 49426) but appears to support good growth of other *Sulfolobus* isolates when appropriate carbon and nitrogen sources are used (Grogan 1989). The complete medium should not be autoclaved; if desired, however, bottled liquid medium can be steamed for 30 minutes, which prevents growth of contaminants during storage at room temperature. Efficient aeration of liquid cultures is necessary to attain high cell densities and can be achieved by continuous agitation or by limiting the depth of static cultures to 2–3 mm. Cell density can be conveniently monitored by photometry.

• The gel formed by 0.7% Gelrite plus Mg^{++} and Ca^{++} is thermostable and can withstand up to several weeks of incubation at 75°C. This property also prevents the gel from being remelted, so take care to pour the plates before the medium cools below approximately 60°C. Some brands of disposable (polystyrene) plates do not withstand incubation conditions for *Sulfolobus*; Becton Dickinson plates (Fisher 08-757-100D) usually hold up well.

Direct Selection of Resistance Mutants

MATERIALS

Liquid culture of wild-type *Sulfolobus* (see pp. 125–126)
Gelrite plates (Kelco division of Merck, distributed by SchweizerHall 46-3070-00) with selective agent (see below)
Glass spreader

METHODS

1. Grow a fresh liquid culture to a final cell density of approximately 3×10^8 cells per milliliter (1–3 days).

Table 1 Spontaneous Resistance Mutants of *S. acidocaldarius* Strain DG6

Inhibitor	Mutant frequency	Minimum inhibitory concentration (μg/ml)	
		parent	mutants
L-Ethionine	2×10^{-9}	200	>2,000
5-Fluorouracil[a]	$1–10 \times 10^{-7}$	15	50–2,000
5-Fluorocytosine[a]	$1–10 \times 10^{-7}$	120	250–2,000
5-Fluorouridine[a]	$1–10 \times 10^{-7}$	260	1,000–10,000
5-Fluoro-orotic acid	$1–10 \times 10^{-7}$	8	60–>1,000
L-Canavanine	1×10^{-8}	≤10	(not determined)

[a]These three compounds apparently select only one class of mutants, designated *fpy* (Grogan and Gunsalus 1993).

2. Spread cells directly on plates containing one selective agent at approximately twice the minimum inhibitory concentration listed in Table 1. For the canavanine and methionine selections, concentrate the cells tenfold so that only small volumes (<0.2 ml) need to be plated.

3. Incubate for 7–12 days.

4. Pick colonies and streak the cells for isolation on nonselective medium.

COMMENTARY

- The compounds listed in Table 1 have been found to inhibit growth of *Sulfolobus* strains and to select spontaneous mutants of strain DG6 (Grogan 1991). With the exception of auxotrophic mutants selected by 5-fluoro-orotate (see below), the precise biochemical bases for the acquired resistances have not been determined.

- Novobiocin has also been observed to select spontaneous resistance mutants (Grogan 1991), but this selection has proven less reliable than listed in Table 1. A mutant's level of resistance to the selective agent can be conveniently measured in liquid medium by inoculation of a dilution series, e.g., in a microdilution plate.

Direct Selection of Pyrimidine Auxotrophs

MATERIALS

Liquid culture of *S. acidocaldarius* (see pp. 125–126)
Gellan gum plates containing uracil and/or 5-fluoro-orotic acid (see below)
Sterile toothpicks or wooden applicator sticks

METHODS

1. Prepare xylose-tryptone plates containing 50 μg/ml fluoro-orotate and 20 μg/ml uracil.

2. Spread $2–5 \times 10^8$ cells per plate and incubate 5–10 days.

3. Score the resulting colonies as follows. Pick individual colonies with a sterile toothpick or the blunt end of a sterile wooden applicator stick. Spot each colony in a grid pattern on a succession of plates supplemented with (*i*) uracil and 50 µg/ml fluoro-orotate, (*ii*) 50 µg/ml fluoro-orotate only (no uracil), and (*iii*) uracil and 300 µg/ml fluoro-orotate.

4. Incubate 3–4 days; *pyrF* mutants will grow on *i* and *iii* but not *ii*, *pyrE* mutants will grow well only on *i*, and *fpy* mutants will grow on *i* and *ii* but not *iii*.

5. Streak mutants of interest for isolation on uracil-containing medium before additional characterization.

COMMENTARY

• In addition to the prototrophic *fpy* class of mutants selected by other 5-fluoropyrimidines (see Table 1), 5-fluoro-orotate selects auxotrophic mutants that are not capable of synthesizing UMP de novo (Grogan and Gunsalus 1993). These require uracil or cytosine, which must be incorporated into the selective medium. One class, designated *pyrE*, has low orotate phosphoribosyl transferase activity, a moderate resistance to fluoro-orotate (MIC = 180 µg/ml), and continues to grow for several generations after removal of pyrimidine supplement, yielding patchy growth on plates lacking uracil. The other class, designated *pyrF*, lacks both orotate phosphoribosyl transferase and orotidylate decarboxylase, has a high resistance to fluoro-orotate (MIC >1000 µg/ml), and exhibits no residual growth in the absence of uracil (Grogan and Gunsalus 1993).

• *S. acidocaldarius* mutants of the *pyrF* class (1) are readily isolated from any fpy^+ strain, (2) can be readily tested for genetic stability (i.e., phenotypic reversion), and (3) result in a stringent nutritional requirement. This combination of properties has considerable potential in the development of genetic techniques for *Sulfolobus*, as illustrated by the sophisticated techniques developed for yeast and other fungi using the analogous selections (Hoekstra et al. 1991).

Isolation of Auxotrophic Mutants by Replica-plating

MATERIALS

Growth medium for *Sulfolobus* (see pp. 125–126)
Centrifuge (Fisher 04-978-59A)
Microcentrifuge tubes (1.5 ml)
Cylindrical replica-plating block (VWR Scientific 25395-380)
Sterile squares (15 x 15 cm) of velveteen fabric
Gelrite plates (Kelco division of Merck, distributed by SchweizerHall 46-3070-00) containing or lacking nutritional supplements (see below)
MNNG (*N*-methyl *N'*-nitro *N*-nitrosoguanidine)
Sulfolobus dilution (Sdil) buffer

2x Basal Medium (see Appendix 2, Medium 9)	500 ml
gelatin	0.1 g
distilled H_2O	500 ml

Final pH is approximately 4.0; sterilize by autoclaving for 15 minutes.

Mutagenesis buffer

2x Basal Medium (see Appendix 2, Medium 9)	100 ml
MES (2-[*N*-morpholino]ethanesulfonic acid)	2 g
distilled H_2O	97 ml

Adjust pH to 5.5 with KOH and sterilize by filtration.

Caution: MNNG is a potent mutagen and cancer suspect agent. Wear gloves and face mask when handling the solid. Spilled or residual material should be destroyed by dilute HCl or by wiping up in methanol-soaked paper towel and incinerating.

METHODS

1. Grow cells to mid exponential phase (~3×10^8 cells/ml).
2. Harvest by centrifugation at 12,000 rpm for 5 minutes and resuspend to a final density of about 10^9 cells per milliliter in mutagenesis buffer.
3. Carefully weigh 5 mg of MNNG into a small vial and dissolve in 1 ml of distilled H_2O.
4. Distribute the cell suspension equally among three or four microcentrifuge tubes; add the MNNG solution to yield a range of final concentrations (e.g., 0.1–0.5 mg/ml).
5. Mix the suspensions and incubate for 30 minutes at 37°C.
6. Harvest the cells by centrifugation at 12,000 rpm for 1–2 minutes.
7. Carefully remove the supernatant with a pipettor and discard into dilute (~0.2 M) HCl.
8. Resuspend the cells in 1 ml of Sdil buffer and repeat steps 6 and 7.
9. Resuspend the cells in 1 ml of growth medium (see pp. 125–126).
10. Dilute and plate mutagenized cells on fully supplemented medium. The extent of mutagenic treatment, dilution factor, and volume plated must be empirically determined to yield a reasonable number (200–300) of colonies per plate.
11. Stop incubation when the colonies become approximately 0.5 mm in diameter.
12. Invert a plate onto a piece of sterile velvet stretched across the top of a cylindrical replica-plating block (Lederberg and Lederberg 1952). Ensure gentle but uniform contact by tapping lightly at several points on the back of the plate. Remove the plate with a careful, vertical motion.
13. Transfer the resulting pattern of colonies to one supplemented plate and one unsupplemented plate by repetition of step 12. Label each plate to facilitate pairing and comparison after incubation.

14. After all plates have been replicated, incubate until the colony patterns are clearly visible (3–5 days).

15. Allow plates to cool to room temperature. Compare the pattern of two corresponding plates by viewing the supplemented plate through the unsupplemented plate over a dark background.

16. Mark possible auxotrophs (those which grow poorly on unsupplemented plates) on the supplemented plate; streak these for isolation on fresh supplemented plates.

17. For each candidate to be further tested, use a well-isolated colony to inoculate a series of liquid cultures that includes fully supplemented as well as unsupplemented media. It is also convenient to test various pools of supplements at this stage (for examples, see p. 209 of Davis et al. 1980).

18. For those mutants in which auxotrophy is confirmed, repeat step 17 with single precursor compounds to identify the specific requirement, keeping in mind that some single mutations confer multiple requirements.

COMMENTARY

• The above method is an adaptation of the classical replica-plating technique of Lederberg and Lederberg (1952). The nutritional supplements used to grow the colonies determine which auxotrophic phenotypes can be recovered. This consideration may also restrict the carbon and/or nitrogen sources used. (For example, tryptone cannot be used as nitrogen source in the isolation of amino acid auxotrophs.) If any of a number of possible auxotrophs is the goal, complex supplements such as yeast extract could, in principle, be used. However, defined mixtures of precursor compounds help ensure that each is provided at an adequate concentration, and facilitate subsequent identification of the specific requirement of each mutant isolated, via omission tests.

• Prior mutagenesis is probably necessary for effective recovery of *Sulfolobus* mutants by replica plating. Ultraviolet (UV) light can be used as an alternative mutagenic treatment for *S. acidocaldarius* (Grogan 1991). The following steps can thus be substituted for steps 2 through 9 in the above protocol.

1. Harvest the cells by centrifugation and resuspend to a density of approximately 2×10^8 cells per milliliter in Sdil buffer.

2. Transfer the suspension to a flat-bottomed dish to yield a depth of about 1 mm; for example, pipette 5 ml into a 9-cm petri dish.

3. Gently agitate the uncovered suspension under a 15-W germicidal lamp. Exposure for 20 seconds at a distance of about 40 cm should kill about 99% of the cells, which is a reasonable dose for effective mutagenesis. For each situation, actual survival and mutagenic efficiency must be confirmed empirically.

- Using UV treatment of DG6 cells, the replica-plating procedure has yielded histidine, tryptophan, isoleucine-valine, methionine, arginine, pyrimidine, glycerol, and riboflavin auxotrophs (D. Grogan, unpubl.). Other mutagens, such as diethylsulfate and ethylmethanesulfonate, may also be effective for *Sulfolobus* spp., but they have not been tested in the replica-plating procedure. When mutagenized cells are plated, extra time must be allowed for colony formation, as mutagenic treatments generally cause a lag in growth. Similarly, before starting a search for auxotrophic mutants, the supplemented plates should be tested to ensure that they support good growth of colonies. Some supplements inhibit growth and may need to be decreased or eliminated.

- In the above procedure, two replicas are compared to each other (as opposed to comparing a replica to the original) to correct for colony enlargement and possible irregularities caused by uneven transfer to the sterile velveteen. It should be noted that incubating plates too long can cause problems in the replica-plating procedure. Large colonies on the original (master) plate yield smeared replicas. Overincubation of the replicas themselves leads to indentation of the plate surface around the site of growth. This causes optical distortion, which interferes with comparison of colony patterns.

REFERENCES

Davis, R.W., D. Botstein, and J.R. Roth. 1980. *Advanced bacterial genetics.* Cold Spring Harbor Laboratory, Cold Spring Harbor, New York.

Grogan, D.W. 1989. Phenotypic characterization of the archaebacterial genus *Sulfolobus:* Comparison of five wild-type strains. *J. Bacteriol.* **171:** 6710–6719.

——. 1991. Selectable mutant phenotypes of the extremely thermophilic archaebacterium *Sulfolobus acidocaldarius. J. Bacteriol.* **173:** 7725–7727.

Grogan, D.W. and R.P. Gunsalus. 1993. *Sulfolobus acidocaldarius* synthesizes UMP via a standard de novo pathway: Results of a biochemical-genetic study. *J. Bacteriol.* **175:** 1500–1507.

Hoekstra, M.F. H.S. Siefert, J. Nickolof, and F. Heffron. 1991. Shuttle mutagenesis: Bacterial transposons for genetic manipulations in yeast. *Methods Enzymol.* **194:** 329–342.

Lederberg, J. and E.M. Lederberg. 1952. Replica plating and indirect selection of bacterial mutants. *J. Bacteriol.* **63:** 399–406.

Miller, J.H. 1972. *Experiments in molecular genetics.* Cold Spring Harbor Laboratory, Cold Spring Harbor, New York.

PROTOCOL 20

Measurement of Mutation Rates in *Sulfolobus acidocaldarius*

D.W. Grogan

Mutational events that enable microorganisms to grow under conditions that normally prevent their growth can be detected even if they occur only once in a billion cell divisions. In principle, if such mutants can be detected and counted in a population, the corresponding mutation rate can be calculated. In practice, however, not every possible selection is useful for this purpose, due to any of a variety of physiological, biochemical, and genetic interferences (Newcombe 1948; Drake 1991). The process of mutation has not been systematically studied in Archaea. Few Archaea have yielded mutant phenotypes, and even fewer (e.g., the extreme halophiles) have demonstrated other genetic phenomena. However, because of the development of convenient plating techniques for *Sulfolobus* spp. (Grogan 1989), isolation of several different classes of mutant strains (Grogan 1991), and biochemical characterization of auxotrophic mutants (Grogan and Gunsalus 1993), at least two classical assays of mutation rate can now be successfully adapted to *Sulfolobus acidocaldarius.*

The first protocol presented below, called the "modified accumulation (A_0) method" (Drake 1991), is a simple colony assay used to measure the rate of forward mutation at the *S. acidocaldarius pyrF* locus, based on the observation that selecting a high level of resistance to 5-fluoro-orotate in a Pyr$^-$ strain of *S. acidocaldarius* will yield only *pyrF* mutants (Grogan and Gunsalus 1993). The second protocol describes a Poisson distribution method using multiple liquid cultures (Ryan 1959) to measure phenotypic reversion of a *pyrF* mutation in *S. acidocaldarius.* This latter method (see Fig. 1) uses nutrient limitation to stop the growth of many independent liquid cultures at a cell density at which about half of the cultures should possess a prototrophic revertant.

Forward Mutation Using the A_0 Method

MATERIALS

Growth medium for *Sulfolobus acidocaldarius* (see Protocol 19)
Incubator (75°C or higher)
Gellan gum plates (Kelco division of Merck, distributed by SchweizerHall 46-3070-00) containing solid medium with 5-fluoro-orotate and uracil (see Protocol 19)
Petroff-Hausser counting chamber or hemacytometer (e.g., Baxter, Scientific Products B3192-1)
Spectrophotometer or colorimeter (e.g., Baxter, Scientific Products C6800)
5-fluoro-orotate (Sigma F 5013)

METHODS

1. Streak DG64 (Grogan and Gunsalus 1993) or other suitable PyrF+ mutant for isolation. All media in this procedure should contain 20 μg of uracil per milliliter to support growth of the auxotrophs selected.

2. When the colonies are still rather small, inoculate a series of tubes (e.g., 10) containing 1 ml of medium, each with a different colony. Incubate without shaking until turbidity can be seen.

3. Add at least 2 ml of fresh medium to each tube and reincubate with aeration. Monitor subsequent growth photometrically.

4. When the cultures reach moderate density but are still growing (approximately 4×10^8 cells/ml), stop incubation and estimate the volume of each culture.

5. Plate aliquots of 50, 100, and 200 μl from each culture on separate plates containing 250 μg/ml 5-fluoro-orotate; incubate for 7–10 days.

6. Determine the density (cells/ml) of the suspensions plated in step 5, either by plating appropriate dilutions on uracil-supplemented plates without selective agent or by comparing the turbidity at the time of plating to a standard curve made by particle and viable counts.

7. Count the resulting colonies resistant to fluoro-orotate. The number of colonies should be roughly proportional to the volume plated. If not, try adjusting the plating or incubation conditions.

8. Divide the average number of mutants per milliliter of culture plated by the cell density. This yields the mutant frequency, f, for that culture.

9. Determine which of the cultures has the median of f for the set. Use the data from this culture to calculate the mutation rate μ by iterative solution of the equation $\mu = f/\ln(N\mu)$ (Drake 1991), as follows:

   a. Calculate the total number of cells, N, in the median culture.
   b. Estimate a value of μ that is about 20% f for this culture.
   c. Calculate the resulting $f/\ln(N\mu)$.
   d. Use this value as a new approximation of μ to repeat step c.
   e. Repeat the iteration until a relatively constant value is reached.

   This is the observed mutation rate, i.e., the number of detectable mutational events per cell division.

COMMENTARY

• Mutations can be distinguished as "forward" or "reverse" according to the biochemical change selected. Forward mutations that can be directly selected in bacteria typically result from loss of an enzyme, binding site, or regulatory mechanism; the loss makes the cell resistant to an inhibitor (Vinopal 1987). In the case above, loss of an enzyme of UMP biosynthesis, orotidine monophosphate decarboxylase, makes the cell resistant to 5-fluoro-orotic acid. Many selections for forward mutation have the advantage of being applicable to the wild-type strain, but they tend to miss recent mutational events because of phenotypic lag, due to the consequence that sensitivity is often "dominant" over resistance (Newcombe 1948). Reverse mutations can be detected in an appropriate mutant strain. This is usually an auxotroph, in which additional mutations can correct the original metabolic defect and restore growth in minimal medium. Such a case is illustrated in the P_0 method (Ryan 1959) described below. Reverse mutations are usually not masked by phenotypic lag, but they may be difficult to define at the molecular level, as they often include suppressor mutations at other loci (second-site revertants).

- The above procedure uses *pyrB* strain DG64 to ensure that no *fpy* mutants are recovered by the selection (Grogan and Gunsalus 1993). This is probably not necessary, however, since most *fpy* mutants are not resistant to the high levels of 5-fluoro-orotic acid used in the above protocol (Grogan and Gunsalus 1993). Although multiple independent cultures are plated in the A_0 procedure above, only one is used to calculate μ, namely, the culture yielding the median number of mutants. This is to reduce the effects of the intrinsic randomness of the mutation process; the median, rather than the average, value was taken in order to prevent the upward bias caused by the few "jack-pot" cultures in which a mutant arose very early (Newcombe 1948).

Reverse Mutation by the P_0 Method

MATERIALS

Growth medium for *Sulfolobus acidocaldarius* (see Protocol 19)
Microdilution plates (96-well; e.g., Fisher 08-772-5)
Spectrophotometer or colorimeter (e.g., Baxter, Scientific Products C6800)
Petroff-Hausser counting chamber or hemacytometer (e.g., Baxter, Scientific Products B3192-1)

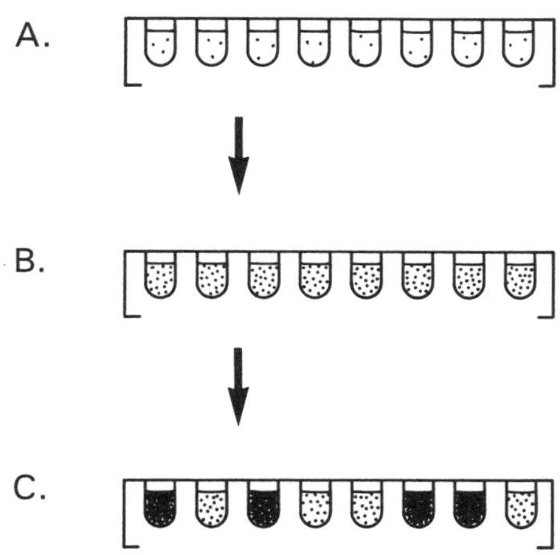

Figure 1 Summary of P_0 Method (see Ryan 1959). (*A*) A series of identical liquid cultures is prepared. Each culture contains a few thousand auxotrophic cells and a limiting amount of the required nutrient. (*B*) Cultures are incubated, and growth continues until the nutrient is exhausted. At this point, each culture has enough cells so that the probability of containing at least one revertant is about 50%. (*C*) Incubation continues until the revertant cells have overgrown the cultures in which they arose.

METHODS

1. Streak a suitable *pyrF* strain for isolation. Stop incubation before the colonies exceed 0.5 mm in diameter.

2. Inoculate 1 ml of xylose-tryptone medium (supplemented with 5 μg of uracil/ml) with a well-isolated colony. Incubate until growth can just be detected (~3×10^7 cells/ml).

3. Make three small flasks of medium (12 ml each) *lacking uracil.* Inoculate each with 50, 100, or 200 μl of the above culture, respectively.

4. Prepare three 96-well microdilution plates by filling the outermost wells with distilled H_2O. This leaves 60 wells empty. Distribute 0.2 ml of each flask into the wells of the corresponding plate. Label the plate with the volume of inoculum used (e.g., 50 μl).

5. Arrange the plates as a stack and place extra, water-filled microdilution plates on top and bottom. Place inside a well-sealed container and incubate. Inspect growth at 5 days and daily thereafter.

6. When a clear difference can be seen between wells that have completely grown and those that have stopped growing, discontinue incubation.

7. For each plate, record the number of wells that failed to grow to saturation. These are the wells that contain no revertants and define the P_0 term of the Poisson distribution. For accurate results, this proportion should be approximately half; if it is very large or very small, the inoculum size (step 3) should be increased or decreased, respectively.

8. Using a pipettor, pool the contents of as many of the above wells as possible from a given plate (i.e., three separate pools). Take care to resuspend and remove cells quantitatively, keeping count of the number of wells pooled.

9. Calculate the total number of cells in each pool by multiplying the volume and the cell density. The cell density can be estimated photometrically, provided the standard curve has been calibrated by microscopic counts of uracil-starved cells, which tend to be larger than normal cells.

10. Determine the average number of cells per well, N_{av}, in the "P_0" term by calculating the total number of cells in each pool and dividing by the number of wells combined.

11. The observed mutation rate (μ) equals $-(\ln 2)(\ln P_0)/N_{av}$.

COMMENTARY

- By itself, tryptone does not support growth of *S. acidocaldarius pyrF* strains and can thus be used in selective media for the P_0 method. This is an advantage because such media support better growth and survival of *Sulfolobus* than other media.

- Although elegant, the P_0 method is limited in that mutation rates must be within a certain range to be measured. If the mutation rate is low (e.g., 10^{-9}), it becomes difficult to distinguish cultures that contain no revertants from those that do. Conversely, mutation rates exceeding 10^{-6} limit the final cell concentrations that can be tolerated in the assay, to the point that N_{av} becomes too small to measure accurately. Thus, some *pyrF* alleles may not work well in this assay. Experimental results for reversion of strain DG100 (*pyrF12*) are approximately 1 x 10^{-7} per cell division (D. Grogan, unpubl.).

REFERENCES

Drake, J.H. 1991. A constant rate of spontaneous mutation in DNA-based microbes. *Proc. Natl. Acad. Sci.* **88:** 7160–7164.

Grogan, D.W. 1989. Phenotypic characterization of the archaebacterial genus *Solfolobus:* Comparison of five wild-type strains. *J. Bacteriol.* **171:** 6710–6719.

———. 1991. Selectable mutant phenotypes of the extremely thermophilic archaebacterium *Solfolobus acidocaldarius. J. Bacteriol.* **173:** 7725–7727.

Grogan, D.W. and R.P. Gunsalus. 1993. *Sulfolobus acidocaldarius* synthesizes UMP via a standard de novo pathway: Results of a biochemical-genetic study. *J. Bacteriol.* **175:** 1500–1507.

Newcombe, H.B. 1948. Delayed phenotypic expression of spontaneous mutations in *E. coli. Genetics* **33:** 447–476.

Ryan, F.J. 1959. Bacterial mutation in stationary phase and the question of cell turnover. *J. Gen. Microbiol.* **21:** 530–549.

Vinopal, R.T. 1987. Selectable phenotypes. In Escherichia coli *and* Salmonella typhimurium: *Cellular and molecular biology* (ed. F.C. Neidhardt et al.), pp. 990–1015. American Society for Microbiology, Washington, D.C.

Typing Marine Vent Thermophiles by DNA Polymerase Restriction Fragment Length Polymorphisms

F.B. Perler, M.W. Southworth, D.G. Wilbur, and D. Wallace

It is often difficult to differentiate or identify marine vent isolates because of the problems in culturing sufficient quantities of cell mass to perform standard typing procedures. This protocol describes a simple and sensitive restriction fragment length polymorphism (RFLP) test to compare marine vent isolates from small cultures. The protocol includes small-scale culture, preparation of DNA, and RFLP analysis of DNA polymerase genes. Proceed to the RFLP section if DNA has already been obtained.

Growth of Marine Vent Thermophiles

MATERIALS

Growth medium (see Protocols 1–4)

Marine Broth 2216 (Difco 0791-01-2) supplemented (per liter) with 1.41 g of 0.005 M Bis-Tris propane and 1.21 g of 0.01 M cysteine. Adjust pH to 8.2 with NaOH. Autoclave to sterilize. If required, add elemental sulfur (8 g/liter) after sterilization.

Inoculum (10 ml) of isolates to be tested (Belkin and Jannasch 1985).

Stoppered bottles (VWR 16159-981). To seal bottles, tighten the cap completely and then loosen one-quarter of a turn to vent.

Spectrophotometer

Cheesecloth prefilter (VWR 21910-130)

Centrifuge (GSA rotor, Beckman or equivalent)

METHODS

1. Inoculate 1 liter of media in a screw-top bottle with a 10-ml starter culture. Incubate the flask in an 85°C oven as a static culture.

2. Monitor growth by visual turbidity and with a spectrophotometer at 550 nm. Grow cultures for 20–68 hours, depending on isolate growth rate, until an optical density of approximately 0.2 at A_{550} (~1×10^8 cells/ml) is reached.

3. Harvest cultures by dripping through a cheesecloth prefilter to remove elemental sulfur. Collect cells by centrifugation at 8000 rpm for 5 minutes in the cold in a Beckman GSA rotor or equivalent.

4. Store frozen cells at –70°C.

COMMENTARY

• With some isolates, elemental sulfur may be eliminated. Test for growth in the absence of elemental sulfur for easier DNA purification.

• This protocol assumes that a seed culture of the unknown isolate has already been obtained. The yield of cells is typically 0.5–1 g/liter (wet weight). This small-scale culture procedure has been successful with many marine vent isolates but may not be applicable to all. Therefore, utilize the best small-scale culture procedure known for the test isolate.

Chromosomal DNA Isolation Procedure

MATERIALS

Bench-top centrifuge
Eppendorf microcentrifuge
Microcentrifuge tubes (1.5 ml)
Gibson Pipetman (Rainin P-20, P-200, P-1000)
Pipette tips
Horizontal electrophoresis apparatus
Power source
Lysozyme, 10 mg/ml in TE buffer
1% SDS (sodium dodecyl sulfate)
Phenol (equilibrated with Tris-HCl, pH 8) (prepared as described by Sambrook et al. 1989)
Chloroform
NaOAc ((3 M, pH 5.5)
Isopropanol
Agarose
EtOH (70%)
RNase (heat-treated; 10 mg/ml) (prepared as described by Sambrook et al. 1989)
1x Lysis buffer (pH 8): 50 mM Tris (pH 8), 50 mM EDTA, 20% sucrose
1x TE Buffer: 10 mM Tris-HCl, 0.1 mM EDTA (pH 7)

Cautions: Chloroform is irritating to the skin, eyes, mucous membranes, and respiratory tract. Work in a chemical fume hood and wear gloves and safety glasses. Chloroform is a carcinogen and may damage the liver and kidneys.

Phenol is highly corrosive and can cause severe burns. Wear gloves, protective clothing, and safety glasses and work in a chemical fume hood. Any areas of skin that come in contact with phenol should be rinsed with a large volume of water or PEG 400 and washed with soap and water; do not use ethanol!

Wear a mask when weighing SDS.

METHODS

1. Thaw pellets at room temperature or 37°C and resuspend in 1.5 ml of lysis buffer.

2. Add 150 μl of lysozyme (10 mg/ml), mix carefully, and let stand at room temperature for 15 minutes.

3. Add 150 μl of 1% SDS, mix carefully, and let stand at room temperature for 15 minutes.

4. Add 1 ml of phenol and mix gently, followed by 1 ml of chloroform. Mix again. Spin in a bench-top centrifuge at maximum speed for 5–10 minutes. Collect upper aqueous layer with a wide-bore pipette tip.

5. Reextract phenol/chloroform layer with 0.5 ml of lysis buffer, spin, and pool aqueous phase with previously collected aqueous phase.

6. Repeat phenol/chloroform extraction on pooled aqueous sample. Repeat four to five times until the sample is clear and there is no interphase.

7. Add an equal volume of chloroform alone for the final extraction, mix, spin, and collect upper aqueous phase.

8. Add 2 μl of RNase (10 mg/ml) to aqueous sample, which should be approximately 2 ml in volume. Incubate 30 minutes at room temperature.

9. Phenol/chloroform extract again.

10. Aliquot 0.5 ml of the DNA solution into microfuge tubes for precipitation. Precipitate DNA with 1/10 volume 3 M NaOAc at pH 5.5 (50 μl) and 1 volume isopropanol (0.5 ml). Mix and incubate 2 hours to overnight at –20°C.

11. Spin samples in a microfuge at maximum speed for 30 minutes at 4°C. Carefully remove supernatant and wash pellet in 70% EtOH. Spin and carefully remove remaining EtOH with a pipettor. Air-dry for 30 minutes.

12. Resuspend DNA in 150 μl of TE buffer. Incubate DNA pellet for 30 minutes at 37°C followed by overnight at 4°C. The pellet can also be mixed by gently pipetting, but take care not to lose the pellet in the pipette tip.

13. Analyze 10 μl of DNA by electrophoresis on a 1% agarose gel with DNA concentration standards to determine quality of DNA (breakdown) and concentration.

14. Dialyze the remainder of the DNA solution extensively against TE buffer (three changes of 1 liter for at least 2 hours each at 4°C).

15. Store at –20°C in aliquots. Keep the freezing-thawing of the sample to a minimum to prevent degradation.

COMMENTARY

• This procedure should result in 20–200 μg of DNA, depending on isolate. More rapid DNA precipitation may be achieved by placing microfuge tubes directly into crushed dry ice for 10 minutes. We suggest precipitating DNA in microfuge tubes because we find it easier to recover DNA from these tubes.

• It is very difficult to resuspend high-molecular-weight DNA after precipitation. Take care not to shear the DNA by rapidly pipetting through a small opening. Cut the end of the pipette tip to enlarge the opening. The DNA solution should be viscous on resuspension. Otherwise, the yield is low or the DNA is degraded.

• Quantitation of DNA on agarose gels is recommended because DNA solutions often contain RNA or other materials that absorb at A_{260} and interfere with quantitation by spectrophotometer.

RFLP Analysis Procedure

This procedure involves digesting DNA samples with one or two restriction endonucleases and then probing Southern blots with either a 1.3-kb fragment from the 5′end or a 0.7-kb fragment from the 3′end of the *Thermococcus litoralis* DNA polymerase gene (Perler et al. 1991) (GenBank accession No. M74198). Of the 14 marine vent samples tested to date, all hybridize to the Vent DNA polymerase probes, and all except one yield different-size bands. The DNA polymerase genes from GB-3a and Deep Vent (GB-D) were both cloned and found to give identical restriction digestion patterns with ten different restriction enzymes. It is assumed that they are duplicate isolates from the same vent. We suggest using both probes and two enzymes to eliminate the possibility that two isolates fortuitously yield the same pattern with one probe or one digest. Isolates yielding RFLP polymorphisms are different, whereas isolates yielding the same patterns are most likely very similar. Isolates with different RFLP patterns may be closely related since some strains contain intervening sequences in their DNA polymerase genes that would yield RFLP differences.

MATERIALS

Microcentrifuge tubes (1.5 ml)
Horizontal electrophoresis apparatus
Power source
Heat-seal plastic bags (Micro Seal Freezer Bags, American Bioanalytical 702289) and Tew Impulse Sealer (e.g., Chiswick 10008)
Water bath (preferably shaking)
Ethidium bromide solution (10 mg/ml)
DNA (1–2 μg, per isolate)
λ*Hin*dIII or appropriate molecular weight marker
Labeled probe: Vent DNA polymerase 5′ and 3′ probes (Perler et al. 1991)
Nitrocellulose
HCl (0.25 M)
Agarose
Agarose gel running buffer

*Bam*HI (5–20 units/ml)
*Bam*HI buffer (10x): 1.5 M NaCl, 0.1 M Tris-HCl (pH 7.9), 0.1 M $MgCl_2$, 10 mM DDT (dithiothreitol), 1 mg/ml BSA (bovine serum albumin)
50x Denhardt's solution

BSA	1 g
Ficol 400	1 g
PVP (polyvinyl pyrrolidone)	1 g

Adjust to 100 ml with deionized H_2O.
*Eco*RI (5–20 units/ml)
*Eco*RI buffer (10x): 0.5 M NaCl, 1 M Tris-HCl (pH 7.5), 0.1 M $MgCl_2$, 0.25% Triton X-100
4x Hybridization buffer

NaCl	87.7 g
Tris-HCl	36.3 g
EDTA	7.5 g
Na-pyrophosphate	2.0 g
NaH_2PO_4	11.4 g
$Na_2HPO_4 \cdot 7H_2O$	32.2 g

Adjust pH to 8 with HCl, which is also necessary for all components to go into solution, and adjust to 500 ml with deionized H_2O.
Hybridization solution

4x Hybridization buffer	1.25 ml
10% SDS	0.05 ml
50x Denhardt's solution	0.15 ml
denatured carrier DNA	10 μg/ml

Denature probe, final concentration at 1–5 x 10^6 cpm/ml. Adjust to 5 ml with deionized H_2O. Denature carrier DNA and probe by placing in a boiling water bath for 5 minutes.
10 M NH_4Acetate: 77.08 g of NH_4-acetate. Adjust volume to 100 ml with deionized H_2O.
Prehybridization solution

4x Hybridization buffer	12.5 ml
10% SDS	0.5 ml
50x Denhardt's solution	10 ml
denatured carrier DNA	10 μg/ml

Adjust to 50 ml with deionized H_2O. Denature carrier DNA by placing in a boiling water bath for 5 minutes.
0.1x SET Hybridization wash buffer (per liter)

20x SET	5 ml
10% SDS	10 ml
10% Na-pyrophosphate	10 ml
1 M $NaPO_4$ (pH 7)	100 ml

20x SET (per liter)

NaCl	175.32 g
Tris base	73 g
EDTA	14.9 g

Southern melt solution: 0.5 M sodium hydroxide, 1 M NaCl
Southern neutralization solution: 1 M Tris-HCl (pH 7.5), 3 M NaCl
Southern transfer buffer: 9 parts 20x SSC, 1 part 10 M NH_4-acetate
20x SSC (per liter): 175.32 g of NaCl, 88.2 g of Na-citrate

Cautions: Ethidium bromide is a powerful mutagen and moderately toxic. Wear gloves when working with solutions that contain this dye. After use, decontaminate and discard according to local safety office regulations.

Wear gloves when handling radioactive substances. Monitor work area throughout. Consult local safety office for further guidance in the appropriate use and disposal of radioactive materials.

METHODS

DNA Digestion

1. Mix the following in a microfuge tube:

   *Eco*RI digestion:

DNA	0.2–1 μg
10x *Eco*RI buffer	3 μl
*Eco*RI (5–20 units)	1 μl

   Add deionized H_2O to 30 μl total volume.

2. Mix the following in a microfuge tube:

   *Bam*HI digestion:

DNA	0.2–1 μg
10x *Bam*HI buffer	3 μl
*Bam*HI (5–20 units)	1 μl

   Add deionized H_2O to 30 μl total volume

3. Incubate the restriction enzyme digestions for 2 hours at 37°C.

4. Load half of each digest on each side of a standard 1% agarose gel along with molecular weight markers. When analyzing more than one DNA sample, group the same enzyme digests for easier comparison. Include 1 ng of the probe on the gel as a positive control and a DNA sample from an organism with a polymerase I type of DNA polymerase as a negative control.

5. Add ethidium bromide (1 μg/ml) in the gel prior to pouring or in the stain for 20 minutes at room temperature after electrophoresis. Photograph the gel with a ruler next to the gel for reference.

Southern Blot Transfer

1. Incubate the gel in 200 ml of 0.25 M HCl for 15 minutes with shaking.

2. Rinse the gel with deionized H_2O and incubate in 200 ml of Southern melt solution for 30 minutes with shaking.

3. Rinse the gel with deionized H_2O and incubate in 200 ml of Southern neutralization solution for 30 minutes with shaking.

4. Transfer the DNA to nitrocellulose at 4°C with Southern transfer buffer for 4 hours to overnight, using standard transfer techniques (Sambrook et al. 1989).

5. After transfer, rinse the filter in 1x SSC for 30 seconds.

6. Air-dry the filter or directly bake in a vacuum oven for 1–2 hours at 60–80°C.

RFLP Hybridization

1. Incubate filter in 10–20 ml of prehybridization buffer for 30 minutes at 50°C.

2. Prepare hybridization buffers.

3. Cut the filter in half to separate duplicate loadings. One half will be probed with the 5′ probe and the other half with the 3′ probe.

4. Blot excess prehybridization buffer from the filter but do not allow to dry.

5. Place each filter in a separate plastic hybridization bag.

6. Add 2–5 ml of hybridization buffer and incubate at 50°C overnight.

7. Carefully remove filters and discard radioactive solution according to local safety office regulations.

8. Wash filters three times for 30 minutes each in 0.1x SET Hybridization wash buffer at 45°C.

9. Wrap the moist filter in plastic wrap and expose to X-ray film.

COMMENTARY

- Prehybridization and hybridization are carried out in sealable plastic bags in shaking water baths. The above protocol describes a moderate stringency hybridization. If no signal is obtained from the unknown isolate, repeat procedure, washing at 37°C instead of 45°C. Film and re-wash again at 42°C.

- This hybridization procedure is designed to reduce background by including nonspecific blocking agents such as Na-pyrophosphate and $NaPO_4$ buffers. Any other hybridization protocol commonly used should substitute well in this procedure, even the use of other types of membranes, if the hybridization stringency and washing temperature are reduced by at least 20%.

- Recommended in this protocol is a 45°C wash temperature because most isolates tested to date cross-hybridize with the Vent DNA polymerase probe at 45°C. This temperature may be tried directly without testing lower stringency wash conditions. It is recommended that the wash temperature be experimentally determined by initially washing at a low temperature such as 37°C, filming, and then increasing the wash temperature by 5°C increments. If the filter is kept moist in plastic wrap during exposure, the optimum wash conditions can be determined. Nonspecific hybridization will wash off with increasing wash temperature until a single or a few bands remain. These specific bands will eventually disappear with increasing wash stringency. The optimum wash temperature is the highest temperature yielding positive bands.

• Best results are obtained when running 0.5–1 μg of DNA per lane, yielding exposure times of as little as 2 hours with a ^{32}P-labeled probe. When using less than 0.5 μg of DNA per lane, expect exposure times of 3–7 days with ^{32}P-labeled probes. We recommend use of at least 0.2 μg of DNA per lane. Exposure times with ^{33}P or nonisotopic probes must be determined experimentally.

• If DNA is resistant to digestion, try adding an extra 10 mM $MgCl_2$. If DNA is limiting, use *Eco*RI instead of *Bam*HI. Alternatively, reprobe the same blot with the second probe or mix both probes together. It is preferable to use bacterial DNA as carrier DNA for hybridizations.

• Probes may be derived from the cloned DNA polymerase gene or from polymerase chain reaction (PCR) of *T. litoralis* DNA (see Genbank, accession No. M74198 for the Vent DNA Polymerase sequence; Perler et al. 1991). The 5′ probe used in the example in Figure 1 contains the coding region, 1–1274 base pairs; the 3′ probe contains the coding region, 4718–5437 base pairs.

• Probes may be labeled by any method (Sambrook et al. 1989). Higher-specific-activity probes allow faster exposure times and ease in rewashing and reexposure. If using radioactive probes, clean to remove unincorporated isotope to improve background and reduce "measles" spots on filters. Elutip-D columns (Schleicher & Schuell 27370) work well.

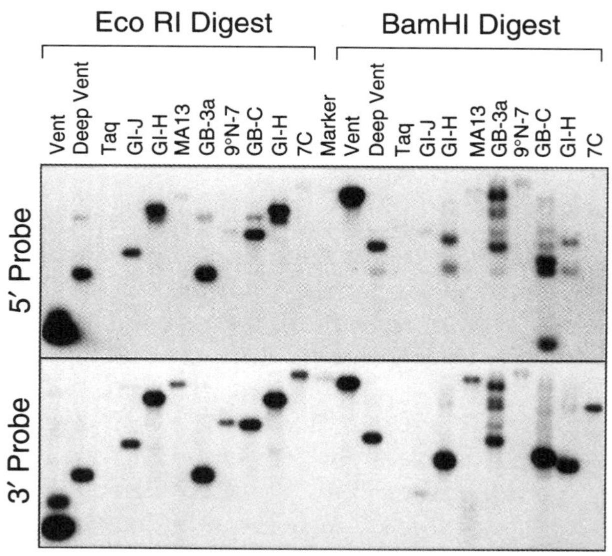

Figure 1 A Typical Blot of Nine Marine Vent Isolates. We used the *T. litoralis* 5′ probe (*top*) and 3′ probe (*bottom*) and DNAs digested with either *Eco*RI (Lanes *1–11*) or *Bam*HI (lanes *13–23*). *T. litoralis* DNA was used as a positive control (lanes *1* and *13*) and *Thermus aquaticus* DNA as the negative control (lanes *3* and *15*). Test strains are GB-D (lanes *2* and *14*), GI-J (lanes *4* and *16*), GI-H (lanes *5* and *17*), MA13 (lanes *6* and *18*), GB-3a (lanes *7* and *19*), 9°N-7 (lanes *8* and *20*), GB-C (lanes *9* and *21*), GI-H (lanes *10* and *22*), and 7C (lanes *11* and *23*). All isolates were supplied by H. Jannasch (Woods Hole Oceanographic Institute, Woods Hole, Massachusetts).

- Fourteen marine vent isolates were tested, and all 14 hybridized positively to the above probes. Samples that did not hybridize and were therefore negative included *Pyrodictium occultum*, *Natronobacterium gregoryi*, *Haloferax volcanii*, *Methanococcus voltae*, and *Sulfolobus acidocaldarius*.

REFERENCES

Belkin, S. and H.W. Jannasch. 1985. A new extremely thermophilic, sulfur-reducing heterotrophic, marine bacterium. *Arch. Microbiol.* **141:** 181–186.

Perler, R.B., D.G. Comb, W.E. Jack, L.S. Moran, B. Qian, R.B. Kucera, J. Benner, B.E. Slatko, D.O. Nwankwo, S.K. Kempstead, C.K.S. Carlow, and H. Jannasch. 1991. Intervening sequences in an Archaea DNA polymerase gene. *Proc. Natl. Acad. Sci.* **89:** 5577–5581.

Sambrook, J., E.F. Fritsch, and T. Maniatis. 1989. *Molecular cloning: A Laboratory Manual*, 2nd edition. Cold Spring Harbor Laboratory Press, Cold Spring Harbor, New York.

PROTOCOL 22

Typing Hyperthermophilic Archaea Based on the 16S/23S rRNA Spacer Region

J. DiRuggiero, J.H. Tuttle, and F.T. Robb

It is difficult to assign new hyperthermophilic isolates to taxa based only on physiological data. We report here a sensitive and convenient method for determining the identity of pure cultures of new isolates. The protocol describes the amplification of the interspace region between the 16S and 23S rRNA and subsequent restriction fragment length polymorphism (RFLP) analysis of the products. If the amplified product contains the endonuclease recognition sequence at unique locations, then the resultant fragment size pattern can be indicative of a particular species (Jensen et al. 1993). The method takes advantage of the fact that the 3′ end of the 16S rRNA gene and the 5′ end of the 23S rRNA are highly conserved, whereas the intervening region is divergent (Vilgalys and Hester 1990; Reysenbach et al. 1992).

MATERIALS

Thermocycler (TEMP-TRONIC; Thermolyne DB66925)
Whatman 3MM filters
Whatman cellulose paper (1001-055)
UV light source
Microcentrifuge tubes (0.5 ml)
CTAB (hexadecyl trimethyl ammonium bromide) (Sigma H 5882)
CTAB/NaCl solution containing 10% (w/v) CTAB and 0.7 M NaCl
Proteinase K (20 mg/ml)
Phenol, TE-saturated (Amresco 0945)
Chloroform/isoamyl alcohol (24:1, v/v)
Isopropyl alcohol
Ethanol (70%, v/v)
NaCl (5 M)
SDS (sodium dodecyl sulfate; 10%, w/v)
Sarkosyl (IBI 07080)
Acetamide (50%, w/v) (Sigma A 0500)
Forward primer 1406F (1391–1406, 16S rRNA *E. coli* numbers), 20 μM
  TGCACACACCGCCCGT
Reverse primer 213aR (228–213, 23S rRNA *E. coli* numbers), 20 μM
  GTTGGTTTCTTTTCCT
dNTPs: 2 mM (each) of dATP, dCTP, dGTP and dTTP in double-distilled H_2O
TE buffer (1x): 0.01 M Tris-HCl, 0.1 mM EDTA (pH 7.6)

Taq buffer (10x): 500 mM KCl, 100 mM Tris-HCl (pH 9.0), Triton X-100, 1% (v/v)
Restriction enzymes: *Taq* polymerase and corresponding buffers were from Promega. All the solutions are prepared with sterile double-distilled H_2O.
Taq DNA polymerase (5000 units/ml)
*Hpa*II (6000 units/ml)
*Hpa*II buffer (10x): 60 mM Tris-HCl (pH 7.5), 60 mM NaCl, 60 mM $MgCl_2$, 60 mM β-mercaptoethanol
*Hha*I (12,000 units/ml)
*Hha*I buffer (10x): 100 mM Tris-HCl (pH 7.9), 500 mM NaCl, 100 mM $MgCl_2$, 10 mM DTT (dithiothreitol)
$MgCl_2$ (25 mM)
BSA (bovine serum albumin; 1 mg/ml)
Wizard PCR Preps purification kit (Promega A7100)
Low-melting agarose (USB 9012-36-6)
Mineral oil
MetaPhor agarose (FMC Bioproducts 50182)
Ethidium bromide solution (1 mg/ml)
PGem molecular weight marker (Promega 275504)
10x TBE (see Ausubel et al. 1987)
10x TAE (see Ausubel et al. 1987)

Cautions: Chloroform is irritating to the skin, eyes, mucous membranes, and respiratory tract. Work in a chemical fume hood and wear gloves and safety glasses. Chloroform is a carcinogen and may damage the liver and kidneys.

Ethidium bromide is a powerful mutagen and moderately toxic. Wear gloves when working with solutions that contain this dye. After use, decontaminate and discard according to local safety office regulations.

β-Mercaptoethanol may be fatal if inhaled or absorbed through the skin and is harmful if swallowed. High concentrations are extremely de-structive to the mucous membranes, the upper respiratory tract, the skin, and the eyes. Wear gloves and safety glasses and work in a chemical fume hood.

Phenol is highly corrosive and can cause severe burns. Wear gloves, protective clothing, and safety glasses and work in a chemical fume hood. Any areas of skin that come in contact with phenol should be rinsed with a large volume of water or PEG 400 and washed with soap and water; do not use ethanol!

Ultraviolet radiation is dangerous, particularly to the eyes. To minimize exposure, make sure that the ultraviolet light source is adequately shielded and wear protective goggles or a full safety mask that efficiently blocks ultraviolet light.

METHODS

DNA Preparation

1. Grow cultures (for standard growth protocol, see Protocol 6) to at least 1 x 10^8 cells per milliliter. Filter the culture through Whatman cellulose paper to remove most of the sulfur before centrifugation.

2. Extract DNA by the CTAB method described by Ausubel et al. (1987) and modify as follows:

Add Sarkosyl (1% final) to the cell lysis mixture and treat the lysate with proteinase K (1 mg/ml final) overnight at 55°C.

After incubation with the CTAB/NaCl solution, extract three times with an equal volume of phenol/chloroform/isoamyl alcohol (24:24:1, v/v/v) and once with chloroform/isoamyl alcohol (24:1, v/v).

PCR Amplification

1. Prepare the following mixture in sterile 0.5-ml microcentrifuge tubes. Keep all reagents and solutions at 4°C.

10x *Taq* buffer	10 μl
$MgCl_2$	10 μl
dNTPs	10 μl
Forward primer	10 μl
Reverse primer	10 μl
Acetamide	10 μl
DNA	100–200 ng
Taq DNA polymerase	10 units
Double-distilled H_2O	to 100 μl

2. Overlay the reaction mixture with 20 μl of mineral oil. To avoid DNA contaminations from buffers, water, or primers, expose the reaction mixture to UV light using a transilluminator for 3 minutes *before adding the DNA and the enzyme.* Prepare a negative control without DNA for each set of reactions. For the amplification of multiple samples, prepare a master mix containing all the reaction components except DNA, and aliquot into individual 0.5-ml microcentrifuge tubes.

3. Thermocycler program

   95°C for 3 minutes, followed by
   30 cycles of
   - 95°C for 1.5 minutes
   - 48°C for 1 minute
   - 72°C for 2 minutes

   Allow the thermocycler to reach 90°C before placing the tubes in the heat block to avoid any priming in the ramp and subsequent primer extension prior to complete DNA denaturation.

4. Analyze the amplified products (20 μl) on a 1.2% agarose gel containing ethidium bromide in 1x TAE buffer. The size of the products is between 500 and 600 base pairs, and a single band was observed for all of the hyperthermophiles examined.

PCR Product Purification

1. Purify the products either directly or from a 1.2% low-melting agarose, using the Wizard PCR Preps purification kit and resuspend in 50 μl of H_2O.

2. If a second round of amplification is necessary to produce sufficient DNA for restriction analysis, use 5 μl of the purified products from the first round of amplification following the same protocol.

3. Estimate the DNA concentration for each product by loading 3–5 μl on a 1.2% agarose gel together with a known concentration of a molecular weight marker.

RFLP Protocol

Double digestions, using *Hpa*II and *Hha*I, are carried out as follows.

1. Prepare the following mixture in sterile 0.5-ml microcentrifuge tubes. Keep all reagents and solutions on ice.

10x enzyme buffer	2.5 μl
BSA	2.5 μl
DNA	500 ng
*Hpa*II	3 units
*Hha*I	6 units
Double-distilled H_2O	to 25 μl

2. Incubate for 1 hour at 37°C. Add 1 μl of 0.5 M EDTA to stop the reaction.

3. Separate digests on a 4% MetaPhor agarose gel without ethidium bromide for 4 hours at 5 V/cm. Before photographing, stain with ethidium bromide for 20 minutes and destain with double-distilled H_2O for 20 minutes, with at least one change. If the DNA fragments do not resolve completely, electrophorese the gel further (1–2 hours) and photograph again without restaining.

COMMENTARY

- The DNAs from several hyperthermophilic Archaea used in this protocol were purified using the modified CTAB method (described in this protocol), which gives good quality DNA and can be performed on a very small scale. Nevertheless, different DNA extraction procedures (see, e.g., Protocol 14) may be required to produce suitable quality DNA from different hyperthermophiles.

- The primers used for the amplification of the spacer region are located within the 16S and 23S rRNAs, in highly conserved regions (for more information about the primers, see Appendix 6). The 16S and 23S rRNAs of thermophilic Archaea are generally transcriptionally coupled (Brown et al. 1989), except for *Thermoplasma acidophilum* (Ree and Zimmermann 1990). In our hands, eight hyperthermophilic Archaea produced single bands when the 16S/23S rRNA interspace region was amplified. This procedure has also been used for bacteria, which can possess several rRNA operons. In this case, the length and sequence polymorphisms present in the PCR product identify genera and species without need for restriction pattern analysis (Jensen et al. 1993).

- Because the area targeted for amplification exhibits a high degree of single-strand secondary structure which can compete with primer hybridization, a longer denaturation time (1.5 minutes) and the addition of acetamide (5% final concentration) were required. Acetamide, a denaturant, has been shown to maximize conditions for template denaturation and primer annealing in PCRs (Reysenbach et al. 1992). For information on the use of acetamide, see Protocol 16.

• Effective purification of the PCR products is critical in order to achieve complete enzyme digestion and clear resolution of fragments. Some of the fragments are very small and the interpretation of the results is difficult when traces of the primers are present in the mixture. The low-melting agarose gel purification results in a better yield than direct purification with the Promega system. Two successive amplification reactions generally produce enough DNA to preclude the gel purification step, even when the starting concentration of genomic DNA is very low.

• The MetaPhor agarose used to analyze the restriction digests results in resolution equivalent to acrylamide gels even for very small fragments. The two enzymes used for the RFLP analysis are 4-base-pair cutters recognizing GC-rich sites and are most likely to cleave a 5–600-base-pair insert. In some cases, a different size insert was obtained by amplification, eliminating the need for RFLP analysis. *Hpa*II treatment resulted in three to five bands and *Hha*I resulted in two to four bands. This permitted discrimination between seven hyperthermophilic Archaea, some of which are very close phylogenetically according to the sequence of the glutamate dehydrogenase gene (DiRuggiero et al. 1993). Two other enzymes, *Alu*I and *Hae*III, were tested. These enzymes did not allow discrimination between the organisms examined.

REFERENCES

Ausubel, F.M., R. Brent, R.E. Kingston, D.D. Moore, J.A. Smith, J.G. Seidman, and K. Struhl. 1987. Preparation and analysis of DNA. In *Current protocols in molecular biology*, pp. 2.4.1–2.4.5. Greene/Wiley, New York.

Brown, J.W., C.J. Daniels, and J.N. Reeve. 1989. Gene structure, organization and expression in archaebacteria. *Crit. Rev. Microbiol.* **16:** 287–338.

DiRuggiero, J., F.T. Robb, R. Jagus, H.H. Klump, K.M. Borges, M. Kessel, X. Mai, and M.W.W. Adams. 1993. Characterization, cloning, and in vitro expression of the extremely thermostable glutamate dehydrogenase from the hyperthermophilic archaeon, ES4. *J. Biol. Chem.* **268:** 17767–17774.

Jensen, M.F., J.A. Webster, and N. Straus. 1993. Rapid identification of bacteria on the basis of polymerase chain reaction-amplified ribosomal DNA spacer polymorphisms. *Appl. Environ. Microbiol.* **59:** 945–952.

Ree, H.K. and R.A. Zimmermann. 1990. Organization and expression of the 16S, 23S and 5S ribosomal RNA genes from the archaebacterium *Thermoplasma acidophilum*. *Nucleic Acids Res.* **18:** 4471–4478.

Reysenbach, A.L., L.J. Giver, G.S. Wickham, and N.R. Pace. 1992. Differential amplification of rRNA genes by polymerase chain reaction. *Appl. Environ. Microbiol.* **58:** 3417–3418.

Vilgalys, R. and M. Hester. 1990. Rapid genetic identification and mapping of enzymatically amplified ribosomal DNA from several *Cryptococcus* species. *J. Bacteriol.* **172:** 4238–4246.

PROTOCOL 23

In Vitro Transcription from Natural and Mutated rDNA Promoters of the Thermophilic Archaeon *Sulfolobus shibatae*

J. Hain and W. Zillig

An in vitro transcription system prepared from *Sulfolobus* (Hüdepohl et al. 1990) was established to determine the basic elements of an archaeal promoter (Reiter et al. 1990) and to dissect these elements further (Hain et al. 1992). The soluble fraction of a cell-free extract of *Sulfolobus shibatae* (Hüdepohl et al. 1990) specifically initiates the transcription of several homologous and heterologous promoters at their respective in vivo start sites (Hüdepohl et al. 1991). S1 analysis of the products showed that it was possible not only to map the start sites of isolated natural or mutant promoters, but also to quantify their efficiency by densitometric analysis of the autoradiograms (Reiter et al. 1990). The templates used were constructed by inserting fragments encompassing whole promoter regions of several archaeal genes into pUC18 (Hüdepohl et al. 1990).

MATERIALS

Prepare all solutions with diethyl pyrocarbonate (DEPC)-treated water.

Sulfolobus growth medium (see Appendix 2, DSM 88 Medium)

Solution 1	
Tris-HCl (pH 8.0)	50 mM
$MgCl_2$	15 mM
EDTA (pH 8.0)	1 mM
DTT (dithiothreitol)	1 mM
Solution 2	
Tris-HCl (pH 8.0)	0.5 M
$MgCl_2$	0.25 M
EDTA	10 mM
DTT	10 mM
Solution 3	
ATP	20 mM
CTP	10 mM
GTP	10 mM
UTP	10 mM
Solution 4	
sodium acetate (pH 5.0)	100 mM
$MgSO_4$	5 mM

Solution 5	
PIPES (pH 7.0)	50 mM
EDTA	5 mM
sodium trichloroacetic acid (TCA)	3 M

Sephadex G-100 (Sigma G-100-120)
Klenow enzyme (Pharmacia 27-0928-01)
Phenol/chloroform/isoamyl alcohol (24:24:1)
Ethanol (absolute and 70%)
S1 nuclease (Pharmacia 27-0920-01)
DNase I (RNase-free; Boehringer Mannheim 776 785)
Liquid paraffin
Distilled H_2O

Cautions: Chloroform is irritating to the skin, eyes, mucous membranes, and respiratory tract. Work in a chemical fume hood and wear gloves and safety glasses. Chloroform is a carcinogen and may damage the liver and kidneys.

DEPC is a potent protein denaturant and is a suspected carcinogen. It should be handled with care. Wear gloves and work in a chemical fume hood. Point the bottle away from you when opening it; internal pressure can lead to splattering!

Phenol is highly corrosive and can cause severe burns. Wear gloves, protective clothing, and safety glasses and work in a chemical fume hood. Any areas of skin that come in contact with phenol should be rinsed with a large volume of water or PEG 400 and washed with soap and water; do not use ethanol!

TCA is highly caustic. Always wear gloves when handling this compound.

METHODS

Preparation of a Soluble Cell-free Extract

1. Grow *S. shibatae* at 80°C in Brock's medium (see Appendix 2, DSM 88 Medium) complemented with 0.2% sucrose to an OD_{600} of 1.2.

2. Wash the cells twice in distilled H_2O and resuspend in three packed-cell volumes of Solution 1. For lysis, add Triton X-100 to a final concentration of 0.1%.

3. Centrifuge at 55,000 rpm for 40 minutes at 4°C, add glycerol to the supernatant to a final concentration of 20%, and store this cell-free extract at –70°C after freezing in liquid nitrogen. Under these conditions, no loss of activity was observed for up to 6 months (Hüdepohl et al. 1990).

Preparation of ^{32}P-labeled S1 Probe

1. Label an oligonucleotide complementary to a sequence of the in vitro RNA located approximately 80–150 base pairs downstream from the initiation site (Fig. 1) by phosphorylation with [γ-^{32}P]ATP.

2. Hybridize the oligonucleotide to the sense strand DNA of the in vitro RNA and perform a primer extension reaction with Klenow enzyme.

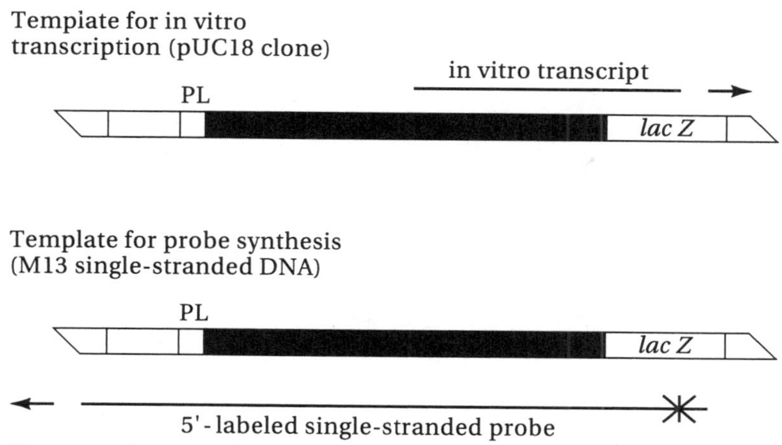

Figure 1 Strategy for S1 Mapping of In Vitro Transcripts. (*Solid bar*) Promoter region of an rRNA gene of *S. shibatae* inserted into the polylinker (PL) region of the respective vector. (*Open bars*) Vector sequences. Asterisk indicates the binding site of the 5′ ^{32}P-labeled M13 universal sequencing primer.

3. Separate the extended primer via Sephadex G-100 chromatography. When analyzing the fractions by Cerenkov-counting in a scintillator, a clear peak of activity should be seen in the elution profile. Pool the fractions containing the first peak, extract them three times with phenol/chloroform/isoamyl alcohol (25:24:1), and store at –20°C.

In Vitro Transcription Reaction

1. Perform standard in vitro transcription reactions for 10 minutes at 60°C in the following mixture:

   0.1–0.3 μg of template DNA (linear or circular)
   5 μl of Solution 2
   5 μl of Solution 3
   8 μl of cell-free extract or 1 μl of purified RNA polymerase (Zillig et al. 1979)
   50 μl of double-distilled H_2O

2. Stop the reactions by chilling on ice and add EDTA to a final concentration of 25 mM. Extract the assay mixtures three times with phenol/chloroform/isoamyl alcohol (25:24:1) and precipitate with ethanol.

3. After centrifugation (13,000 rpm for 25 minutes), washing, and drying, dissolve the pellets in 50 μl of Solution 4 containing 24 units of RNase-free DNase I and incubate for 30 minutes at 25°C.

4. Stop the reactions by adding EDTA to a final concentration of 25 mM and extract three times with phenol/chloroform/isoamyl alcohol (25:24:1). Store the aqueous phase at –80°C.

S1 Nuclease Analysis (Fig. 1)

1. Mix 2 μl (or more) of the in vitro RNA with 10 μg of *Escherichia coli* tRNA, 15,000–50,000 cpm of S1 DNA probe, and precipitate with ethanol.

2. After centrifugation (13,000 rpm for 25 minutes) and washing, dry the pellets totally and resuspend them in 15 µl of Solution 5, add 60 µl of liquid paraffin, heat for 5 minutes at 65°C, and incubate overnight at 45°C.

3. Perform S1 digestion according to published procedures (Murray 1986; Aldea et al. 1988) using 300 units of S1 nuclease per milliliter. The protected fragments are usually analyzed on a 6% polyacrylamide sequencing gel.

COMMENTARY

• It is important to use RNase-free solutions at all steps of the preparation and to avoid contamination with RNases.

• In addition to the *S. shibatae* cell-free extract or purified RNA polymerase (Zillig et al. 1979, 1980), we also tested cell-free extracts of *Sulfolobus solfataricus* and *Sulfolobus acidocaldarius*. The extract of *S. solfataricus* performed similarly to that of *S. shibatae* (T. Singer, unpubl.), but the extract of *S. acidocaldarius* was less efficient due to an endonuclease activity (Prangishvili et al. 1985; P. McWilliams, pers. comm.). The purified RNA polymerase of *S. acidocaldarius* closely resembled the *S. shibatae* enzyme (Hüdepohl 1992).

• Templates containing heterologous promoters from genes of *Halobacterium salinarium*, *Halobacterium mediterranei*, and the *Halobacterium* phage ΦH were examined in the *S. shibatae* transcription system. Not every promoter served for initiation, but if initiation occurred, the major in vitro start site was the same as in vivo (Hüdepohl et al. 1991).

• Specific initiation with the cell-free extract of *S. shibatae* could also be demonstrated in run off transcription experiments.

REFERENCES

Aldea, M., F. Claverie-Martin, M.R. Diaz-Torres, and S.R. Kushner. 1988. Transcript mapping using [^{35}S]DNA probes, trichloroacetate solvent and dideoxy sequencing ladders: A rapid method for identification of transcriptional start sites. *Gene* **65:** 101–110.

Brock, T.D., K.M. Brock, R.T. Belly, and R.L. Weiss. 1972. *Sulfolobus:* A new genus of sulfur-oxidizing bacteria living at low pH and high temperature. *Arch. Microbiol.* **84:** 54–68.

Hain, J., W.-D. Reiter, U. Hüdepohl, and W. Zillig. 1992. Elements of an archaeal promoter defined by mutational analysis. *Nucleic Acids Res.* **20:** 5423–5428.

Hüdepohl, U. 1992. "Die Initiation der Transkription in Archaea." PhD thesis, University of Munich, Germany.

Hüdepohl, U., W.-D. Reiter, and W. Zillig. 1990. In vitro transcription of two rRNA genes of the archaebacterium *Sulfolobus* sp. B12 indicates a factor requirement for specific initiation. *Proc. Natl. Acad. Sci.* **87:** 5851–5855.

Hüdepohl, U., F. Gropp, M. Horne, and W. Zillig. 1991. Heterologous in vitro transcription from two archaebacterial promoters. *FEBS Lett.* **285:** 257–259.

Murray, M.G. 1986. Use of sodium trichloroacetate and mung bean nuclease to increase sensitivity and precision during transcript mapping. *Anal. Biochem.* **158:** 165–170.

Prangishvili, D., R.P. Vashakidze, M.G. Chelidze, and I. Yu. Gabriadze. 1985. A restriction endonuclease *Sua*I from the thermoacidophilic archaebacterium *Sulfolobus acidocaldarius. FEBS Lett.* **192:** 57–60.

Reiter, W.-D., U. Hüdepohl, and W. Zillig. 1990. Mutational analysis of an archaebacterial promoter: Essential role of a TATA box for transcription efficiency and start-site selection in vitro. *Proc. Natl. Acad. Sci.* **87:** 9509–9513.

Zillig, W., K.O. Stetter, and D. Janekovic. 1979. DNA-dependent RNA polymerase from the archaebacterium *Sulfolobus acidocaldarius. Eur. J. Biochem.* **96:** 597–604.

Zillig, W., K.O. Stetter, S. Wunderl, W. Schulz, H. Priess, and I. Scholz. 1980. the *Sulfolobus*-"Caldariella" group. Taxonomy on the basis of the structure of DNA-dependent RNA polymerases. *Arch. Microbiol.* **125:** 259–269.

Appendix 1

Thermophilic Strains Available from DSM*

H. Hippe, B.J. Tindall, and M. Kracht

ACIDIANUS SPECIES

Acidianus brierleyi
Basonym: *Sulfolobus brierleyi*

1651[T] ←W. Zillig and K.O. Stetter←C.L. and J.A. Brierley. Thermal spring drainage; USA, Yellowstone Natl. Park (6). Type strain (6, 7, 40). Taxonomy/description (5). (Medium 150, 70°C)

6334 ←A. Segerer←G. Huber and K.O. Stetter, Sp3a/1. Hot spring drainage; Italy, Pisciarelli. (Medium 358 modified, 75°C, facultatively anaerobic)

Acidianus infernus

3191[T] ←K.O. Stetter, So4a. Mud, solfatara crater; Italy (24). Type strain (40). Taxonomy/description (24). (Medium 358, 88°C, facultatively anaerobic)

ARCHAEOGLOBUS SPECIES

Archaeoglobus fulgidus

4139 ←G. Zellner, strain Z. Marine hydrothermal sediment; Italy, Vulcano Island (41). Taxonomy/description (41). Metabolism (41). S-layer, glycoproteins, cellular polyamines, aphidicolin sensitivity (41). (Medium 399, 78°C, anaerobic)

4304[T] = ATCC 49558
←K.O. Stetter, VC-16. Submarine hot spring; Italy, Vulcano Island (31, 32). Type strain (31, 38). Taxonomy/description (31, 32). Phylogeny (33). (Medium 399, 85°C, anaerobic)

Archaeoglobus profundus

5631[T] ←K.O. Stetter and S. Burggraf, Av18. Deep sea hydrothermal vents; Mexico, Gulf of California, Guaymas Basin (64). Type strain (45). Taxonomy/description (64). (Medium 519, 85°C, anaerobic)

DESULFUROCOCCUS SPECIES

"Desulfurococcus amylolyticus"

3822[T] ←E. Bonch-Osmolovskaya, Z-533. Hot spring; USSR, Kamchatka (43). Proposed type strain (43). (Medium 395, 90°C, anaerobic)

A superscript T after the DSM strain number indicates the strain in the type strain of the species or subspecies.
*DSM is the Deutsche Sammlung von Mikroorganismen und Zellkulturen GmbH, D-38124 Braunschweig, Germany.

Desulfurococcus mobilis
2161^T = ATCC 35582
←W. Zillig, Hvv 3/9. Acidic water from solfataric field; Iceland (13). Type strain (16). (Medium 184, 85°C, anaerobic)

Desulfurococcus mucosus
2162^T = ATCC 35584
←W. Zillig, 07/1. Hot solfataric spring; Iceland (13). Type strain (16). (Medium 184, 85°C, anaerobic)

DESULFUROLOBUS SPECIES

Desulfurolobus ambivalens
3772^T ←W. Zillig, LEI 10. Solfataric mud; Iceland (39). Type strain (27). (Medium 358, 80°C, facultatively anaerobic)

HYPERTHERMUS SPECIES

Hyperthermus butylicus
5456^T ←W. Zillig, PLM1-5. Solfataric sea floor sediment; Azores, San Miguel Island (54). Type strain (55). Taxonomy/description (54). (Medium 491, 99°C, anaerobic)

METALLOSPHAERA SPECIES

Metallosphaera sedula
5348^T ←K.O. Stetter and G. Huber, TH2. Hot water pond; Italy, Naples, Pisciarelli Solfatara (48). Type strain (44). Taxonomy/description (48). (Medium 485, 65°C)

PYROBACULUM SPECIES

"Pyrobaculum aerophilum"
7523^T ←K.O. Stetter, IM2. Hot marine water; Italy, Ischia, Maronti Beach (68). Proposed type strain (68). Taxonomy/ description (68). (Medium 611, 98°C, facultatively aerobic)

Pyrobaculum islandicum
4184^T ←K.O. Stetter and R. Huber, GEO 3. Water, geothermal power plant; Iceland (37). Type strain (29). Taxonomy/description (37). (Medium 390, 95–100°C, anaerobic)

Pyrobaculum organotrophum
4185^T ←K.O. Stetter and R. Huber, H 10. Solfataric spring; Iceland (37). Type strain (29). Taxonomy/description (37). (Medium 390, 95–100°C, anaerobic)

PYROCOCCUS SPECIES

Pyrococcus furiosus
3638^T = ATCC 43587
←K.O. Stetter and G. Fiala, Vc 1. Geothermally heated marine sediment; Italy, Vulcano Island (34). Type strain (26). Taxonomy/description (34). Produces DNA polymerase *Pfu* (67). (Medium 377, 97–100°C, anaerobic)

Pyrococcus woesei
3773^{T} ←W. Zillig, Vul4. Marine solfatara; Italy, Vulcano Island (36). Type strain (28). Taxonomy/description (36). (Medium 377, 97–100°C, anaerobic)

PYRODICTIUM SPECIES

Pyrodictium abyssi
6158^{T} ←K.O. Stetter, AV2. Black smoker at 2011 m depth; Mexico, Gulf of California (62). Type strain (47). Taxonomy/description (62). (Medium 508, 98°C, anaerobic)

Pyrodictium brockii
2708^{T} ←K.O. Stetter, S1. Submarine solfataric field; Italy, Vulcano Island (20). Type strain (23). Taxonomy/description (20). (Medium 283, 85–105°C, anaerobic)

Pyrodictium occultum
2709^{T} ←K.O. Stetter, PL-19. Submarine solfataric field; Italy, Vulcano Island (20). Type strain (23). Taxonomy/description (20). (21, 22). (Medium 283, 85–105°C, anaerobic)

STAPHYLOTHERMUS SPECIES

Staphylothermus marinus
3639^{T} = ATCC 49053
←K.O. Stetter and G. Fiala, F1. Geothermally heated marine sediment; Italy, Vulcano Island (35). Type strain (26). Taxonomy/description (35). (Medium 377, 85–90°C, anaerobic)

3666 ←K.O. Stetter and G. Fiala, AI2. Submarine hydrothermal vent ("black smoker"); East Pacific Rise (35). Taxonomy/description (35). (Medium 377, 85–90°C, anaerobic)

STYGIOLOBUS SPECIES

Stygiolobus azoricus
6296^{T} ←A. Segerer and K.O. Stetter, FC6. Hot spring; Azores, Sao Miguel Island (63). Type strain (63). Taxonomy/description (63). (Medium 510, 80°C, anaerobic)

SULFOLOBUS SPECIES

Sulfolobus acidocaldarius
639^{T} = ATCC 33909, NCIB 11770
←T.D. Brock, 98-3. Acid hot spring; USA, Yellowstone Natl. Park (1). Type strain (4). Taxonomy/description (5). Oxidation of elemental sulfur (3); capability, however, lost (60). (Medium 88, 70°C)

Sulfolobus metallicus
6482^{T} ←K.O. Stetter and G. Huber, Kra 23. Solfataric field; Iceland (65). Type strain (66). Taxonomy/description (65). Ore leaching (65). (Medium 88 with 0.05% elemental sulfur and only 0.02% yeast extract, 65°C)

Sulfolobus shibatae
5389^{T} ←D.W. Grogan←W. Zillig, B12 (*Sulfolobus acidocaldarius*). Geothermal mud hole; Japan, Kiushu Island (50, 51). Type strain (46). Taxonomy/description (49, 51). Virus-like particle (52, 53). Plasmid (50). (Medium 88 with a pH of 3–4, 75–80°C)

Sulfolobus solfataricus

1616^T ←Replacement from W. Zillig (←W. Zillig, P1). Volcanic hot spring; Italy, Campi flegrei (5). Type strain (5, 7). Taxonomy/description (5). [Comment: cultures of DSM 1616 supplied until April 1989 were contaminated with *Sulfolobus acidocaldarius*]. (Medium 182, 70–85°C)

1617 = ATCC 35092
←W. Zillig, P2. Volcanic hot spring; Italy, Campi flegrei (5). (Medium 182, 70–85°C)

5354 ←D.W. Grogan, P1-G←W. Zillig, P1. Volcanic hot spring; Italy, Campi flegrei (5). Taxonomy/description (49). Clonically purified strain derived from strain P1 (49). (Medium 182, 70–85°C)

5833 ←A. Gambacorta←A. Gambacorta and M. De Rosa, MT-4 ("*Caldariella acidophila*"). Solfatara; Italy, Naples, Agnano, Pisciarelli (2, 56). Taxonomy/description (2, 49, 56). Produces alcohol dehydrogenase, aspartate aminotransferase, β-galactosidase, glucose dehydrogenase, malic enzyme, propylamine transferase (60). 2-keto-3-deoxygluconate production (59). Glucose production (58). Metabolism of glucose (76). Trehalose synthesis (57, 61). (Medium 182, 85°C)

THERMOCOCCUS SPECIES

***Thermococcus* sp.**

2770 ←H. Morgan, AN1. Anaerobic mud from alkaline hot spring (25). Taxonomy/description (75). Phylogenetic position (36). (Medium 376, 75°C, anaerobic)

Thermococcus celer

2476^T ←W. Zillig, Vu 13. Solfataric marine water hole (18). Type strain (17). Taxonomy/description (18). Anaerobic sulfur respiration. (Medium 266, 88°C, anaerobic)

"Thermococcus litoralis"

5473^T ←A. Neuner←K.O. Stetter←S. Belkin and H.W. Jannasch, NS-C. Shallow marine thermal spring, beach at Lucrino, near Naples (69). Proposed type strain (73). Taxonomy/description (73). (Medium 623, 83°C, anaerobic)

5474 ←A. Neuner, A3. Shallow marine thermal spring, Vulcano Island (73). Taxonomy/description (73). (Medium 623, 83°C)

Thermococcus stetteri

5262^T ←E.A. Bonch-Osmolovskaya, K-3. Marine volcanic crater fields of Kraternya cove, Ushishir archipelago, Northern Kurils (70). Type strain (45). Taxonomy/description (70). (Medium 480b, 75°C, anaerobic)

THERMOFILUM SPECIES

Thermofilum pendens

2475^T ←W. Zillig, Hrk 5. Solfataric hot spring (19). Type strain (17). Taxonomy/description (19). Nutrition (19). Requires polar lipid fraction from *Thermoproteus tenax* DSM 2078 or from any other archaeon as growth factor (19). Anaerobic sulfur respiration. (Medium 265, 88°C, anaerobic)

THERMOPLASMA SPECIES

Thermoplasma acidophilum

1728^T = AMRC C165, ATCC 25905
←W. Zillig←E.A. Freundt, AMRC C165←ATCC←R.G. Wittler, 122-1B2←T.D. Brock←G. Darland. Coal refuse pile (8, 9). Type strain (4). Taxonomy/description (8, 9, 10, 11). Produces restriction endonuclease *Tha*I (15). (Medium 158, 55–60°C)

Thermoplasma volcanium

4299^T ←A. Segerer and K.O. Stetter, GSS 1. Acid continental solfatara (30). Type strain (38). Taxonomy/description (30). (Medium 398, 60°C)

4300 ←A. Segerer and K.O. Stetter, KD 3. Acid continental solfatara (30). Taxonomy/description (30). (Medium 398, 60°C)

4301 ←A. Segerer and K.O. Stetter, KO 2. Acid continental solfatara (30). Taxonomy/description (30). (Medium 398, 60°C)

THERMOPROTEUS SPECIES

Thermoproteus neutrophilus

2338^T ←K.O. Stetter, V24Sta. Hot spring. Type strain (42). Taxonomy/description (22). (42). (Medium 291, 85°C, anaerobic)

Thermoproteus tenax

2078^T = ATCC 35583
←W. Zillig, Kra 1. Mud hole, solfataric field (12). Type strain (12, 14). Taxonomy/description (12). (Medium 185, 85°C, anaerobic)

"Thermoproteus uzoniensis"

5263^T ←E.A. Bonch-Osmolovskaya, Z-605. Hot springs and soil of the Uzon caldera, SW of the Kamchatka Peninsula (72). Proposed type strain (72). Taxonomy/description (72). (Medium 480c, 85°C, anaerobic)

DSM REFERENCES

1 Brock, T.D., K.M. Brock, R.T. Belly, and R.L. Weiss. 1972. *Sulfolobus*: A new genus of sulfur-oxidizing bacteria living at low pH and high temperature. *Arch. Mikrobiol.* **84:** 54–68.

2 De Rosa, M., A. Gambacorta, G. Millonig, and J.D. Bu'Lock. 1974. Convergent characters of extremely thermophilic acidophilic bacteria. *Experientia* **30:** 866–868.

3 Shivers, D.W. and T.D. Brock. 1973. Oxidation of elemental sulfur by *Sulfolobus acidocaldarius*. *J. Bacteriol.* **114:** 706–710.

4 Skerman, V.B.D., V. McGowan, and P.H.A. Sneath, eds. 1980. Approved lists of bacterial names. *Int. J. Syst. Bacteriol.* **30:** 225–420.

5 Zillig, W., K.O. Stetter, S. Wunderl, W. Schulz, H. Priess, and J. Scholz. 1980. The *Sulfolobus*-*"Caldariella"* group: Taxonomy on the basis of the structure of DNA-dependent RNA polymerases. *Arch. Microbiol.* **125:** 259–269.

6 Brierley, C.L. and J.A. Brierley. 1973. A chemoautotrophic and thermophilic microorganism isolated from an acid hot spring. *Can. J. Microbiol.* **19:** 183–188.

7 Validation of the publication of new names and new combinations previously effectively published outside the IJSB. List No. 5. 1980. *Int. J. Syst. Bacteriol.* **30:** 676–677.

8 Darland, G., T.D. Brock, W. Samsonoff, and S.F. Conti. 1970. A thermophilic, acidophilic mycoplasma isolated from a coal refuse pile. *Science* **170:** 1416–1418.

9 Brock, T.D. 1978. *Thermophilic microorganisms and life at high temperatures*, pp. 92–116. Springer-Verlag, New York.

10 Searcy, D.G. and E.K. Doyle. 1975. Characterization of *Thermoplasma acidophilum* deoxyribonucleic acid. *Int. J. Syst. Bacteriol.* **25:** 286–289.

11 Christiansen, C., E.A. Freundt, and F.T. Black. 1975. Genome size and deoxyribonucleic acid

base composition of *Thermoplasma acidophilum. Int. J. Syst. Bacteriol.* **25:** 99–101.

12 Zillig, W., K.O. Stetter, W. Schäfer, D. Janekovic, S. Wunderl, J. Holz, and P. Palm. 1981. *Thermoproteales*: A novel type of extremely thermoacidophilic anaerobic archaebacteria isolated from Icelandic solfataras. *Zentralbl. Bakteriol. Hyg. I. Abt. Orig. C* **2:** 205–227.

13 Zillig, W., K.O. Stetter, D. Prangishvilli, W. Schäfer, S. Wunderl, D. Janekovic, I. Holz, and P. Palm. 1982. *Desulfurococcaceae*, the second family of the extremely thermophilic, anaerobic, sulfur-respiring *Thermoproteales. Zentralbl. Bakteriol. Hyg. I. Abt. Orig. C* **3:** 304–307.

14 Validation of the publication of new names and new combinations previously effectively published outside the IJSB. List No. 8. 1982. *Int. J. Syst. Bacteriol.* **32:** 266–268.

15 Roberts, R.J. 1982. Restriction and modification enzymes and their recognition of sequences. *Nucleic Acids Res.* **10:** 117–114.

16 Validation of the publication of new names and new combinations previously effectively published outside the IJSB. List No. 10. 1983. *Int. J. Syst. Bacteriol.* **33:** 438–440.

17 Validation of the publication of new names and new combinations previously effectively published outside the IJSB. List No. 11. 1983. *Int. J. Syst. Bacteriol.* **33:** 672–674.

18 Zillig, W., I. Holz, D. Janekovic, W. Schäfer, and W.D. Reiter. 1983. The archaebacterium *Thermococcus celer* represents a novel genus within the thermophilic branch of the archaebacteria. *Syst. Appl. Microbiol.* **4:** 88–94.

19 Zillig, W., A. Gierl, G. Schreiber, S. Wunderl, D. Janekovic, K.O. Stetter, and H.P. Klenk. 1983. The archaebacterium *Thermofilum pendens* represents a novel genus of the thermophilic, anaerobic sulfur respiring *Thermoproteales. Syst. Appl. Microbiol.* **4:** 79–87.

20 Stetter, K.O., H. König, and E. Stackebrandt. 1983. *Pyrodictium*, a new genus of submarine disc-shaped sulfur reducing archaebacteria growing optimally at 105°C. *Syst. Appl. Microbiol.* **4:** 535–551.

21 Stetter, K.O. 1982. Ultrathin mycelia-forming organisms from submarine volcanic areas having an optimum growth temperature of 105°C. *Nature* **300:** 258–260.

22 Fischer, F., W. Zillig, K.O. Stetter, and G. Schreiber. 1983. Chemoautotrophic metabolism of anaerobic extremely thermophilic archaebacteria. *Nature* **301:** 511–513.

23 Validation of the publication of new names and new combinations previously effectively published outside the IJSB. List No. 14. 1984. *Int. J. Syst. Bacteriol.* **34:** 270–271.

24 Segerer, A., K.O. Stetter, and F. Klink. 1985. Two contrary modes of lithotrophy in the same archaebacterium. *Nature* **313:** 787–789.

25 Morgan, H.W. and R.M. Daniel. 1982. Isolation of a new species of sulphur reducing extreme thermophile. Abstr. P 20: 1. XIII. Int. Congress Microbiology, Boston, USA.

26 Validation of the publication of new names and new combinations previously effectively published outside the IJSB. List No. 22. 1986. *Int. J. Syst. Bacteriol.* **36:** 573–576.

27 Validation of the publication of new names and new combinations previously effectively published outside the IJSB. List No. 23. 1987. *Int. J. Syst. Bacteriol.* **37:** 179–180.

28 Validation of the publication of new names and new combinations previously effectively published outside the IJSB. List No. 24. 1988. *Int. J. Syst. Bacteriol.* **38:** 136–137.

29 Validation of the publication of new names and new combinations previously effectively published outside the IJSB. List No. 25. 1988. *Int. J. Syst. Bacteriol.* **38:** 220–222.

30 Segerer, A., T.A. Langworthy, and K.O. Stetter. 1988. *Thermoplasma acidophilum* and *Thermoplasma volcanium* sp. nov. from solfatara fields. *Syst. Appl. Microbiol.* **10:** 161–171.

31 Stetter, K.O. 1988. *Archaeoglobus fulgidus* gen. nov., sp. nov.: A new taxon of extremely thermophilic archaebacteria. *Syst. Appl. Microbiol.* **10:** 172–173.

32 Stetter, K.O., G. Lauerer, M. Thomm, and A. Neuner. 1987. Isolation of extremely thermophilic sulfate reducers: Evidence for a novel branch of archaebacteria. *Science* **236:** 822–824.

33 Achenbach-Richter, L., K.O. Stetter, and C.R. Woese. 1987. A possible biochemical missing link among archaebacteria. *Nature* **237:** 348–349.

34 Fiala, G. and K.O. Stetter. 1986. *Pyrococcus furiosus* sp. nov. represents a novel genus of marine heterotrophic archaebacteria growing optimally at 100°C. *Arch. Microbiol.* **145:** 56–61.

35 Fiala, G., K.O. Stetter, H.W. Jannasch, T.A. Langworthy, and J. Madon. 1986. *Staphylothermus marinus* sp. nov. represents a novel genus of extremely thermophilic submarine heterotrophic archaebacteria growing up to 98°C. *Syst. Appl. Microbiol.* **8:** 106–113.

36 Zillig, W., I. Holz, H.P. Klenk, J. Trent, S. Wunderl, D. Janekovic, E. Insel, and B. Haas. 1987. *Pyrococcus woesei*, sp. nov., an ultra-thermophilic marine archaebacterium, representing a novel order, *Thermococcales. Syst. Appl. Microbiol.* **9:** 62–70.

37 Huber, R., J.K. Kristjansson, and K.O. Stetter. 1987. *Pyrobaculum* gen. nov., a new genus of neutrophilic, rod-shaped archaebacteria from continental solfataras growing optimally at 100°C. *Arch. Microbiol.* **149:** 95–101.

38 Validation of the publication of new names and

new combinations previously effectively published outside the IJSB. List No. 26. 1988. *Int. J. Syst. Bacteriol.* **38:** 328–329.

39 Zillig, W., S. Yeats, J. Holz, A. Böck, M. Rettenberger, F. Gopp, and G. Simon. 1986. *Desulfolobus ambivalens*, gen. nov., sp. nov., an autotrophic archaebacterium facultatively oxidizing or reducing sulfur. *Syst. Appl. Microbiol.* **8:** 197–203.

40 Segerer, A., A. Neuner, J.K. Kristjansson, and K.O. Stetter. 1986. *Acidianus infernus* gen. nov., sp. nov., and *Acidianus brierleyi* comb. nov.: Facultatively aerobic, extremely acidophilic thermophilic sulfur-metabolizing archaebacteria. *Int. J. Syst. Bacteriol.* **36:** 559–564.

41 Zellner, G., E. Stackebrandt, H. Kneifel, P. Messner, U.B. Sleytr, E. Conway de Macario, H.-P. Zabel, K.O. Stetter, and J. Winter. 1989. Isolation and characterization of a thermophilic achaebacterium *Archaeoglobus fulgidus* strain Z. *Syst. Appl. Microbiol.* **11:** 151–160.

42 Schäfer, A., C. Barkowski, and G. Fuchs. 1986. Carbon assimilation by the autotrophic thermophilic archaebacterium *Thermoproteus neutrophilus*. *Arch. Microbiol.* **146:** 301–308.

43 Bonch-Osmolovskaya, E.A., A.I. Sleserev, M.L. Miroshnichenko, T.P. Svetlichnaya, and V.A. Alekseev. 1988. Characteristics of *Desulfurococcus amylolyticus* n. sp.—A new extremely thermophilic archaebacterium isolated from thermal springs of Kamchatka and Kunashir Island. *Microbiology* **57:** 78–84. (Engl. Transl. of Mikrobiologiya)

44 Validation of the publication of new names and new combinations previously effectively published outside the IJSB. List No. 31. 1989. *Int. J. Syst. Bacteriol.* **39:** 495–497.

45 Validation of the publication of new names and new combinations previously effectively published outside the IJSB. List No. 34. 1990. *Int. J. Syst. Bacteriol.* **40:** 320–321.

46 Validation of the publication of new names and new combinations previously effectively published outside the IJSB. List No. 38. 1991. *Int. J. Syst. Bacteriol.* **41:** 456–457.

47 Validation of the publication of new names and new combinations previously effectively published outside the IJSB. List No. 39. 1991. *Int. J. Syst. Bacteriol.* **41:** 580–581.

48 Huber, G., C. Spinnler, A. Gambacorta, and K.O. Stetter. 1989. *Metallosphaera sedula* gen. and sp. nov. represents a new genus of aerobic, metal-mobilizing, thermoacidophilic archaebacteria. *Syst. Appl. Microbiol.* **12:** 38–47.

49 Grogan, D.W. 1989. Phenotypic characterization of the archaebacterial genus *Sulfolobus*: Comparison of five wild-type strains. *J. Bacteriol.* **171:** 6710–6719.

50 Yeats, S., P. McWilliam, and W. Zillig. 1982. A plasmid in the archaebacterium *Sulfolobus acidocaldarius*. *EMBO J.* **1:** 1035–1038.

51 Grogan, D., P. Palm, and W. Zillig. 1990. Isolate B12, which harbours a virus-like element, represents a new species of the archaebacterial genus *Sulfolobus*, *Sulfolobus shibatae*, sp. nov. *Arch. Microbiol.* **154:** 594–599.

52 Martin, A., S. Yeats, D. Janekovic, W.D. Reiter, W. Aicher, and W. Zillig. 1984. SAV1, a temperate, UV-inducible DNA virus-like particle from the archaebacterium *Sulfolobus acidocaldarius* isolate B12. *EMBO J.* **3:** 2165–2168.

53 Zillig, W., F. Gropp, A. Henschen, H. Neumann, P. Palm, W.D. Reiter, M. Rettenberger, H. Schnabel, and S. Yeats. 1986. Archaebacterial virus host systems. *Syst. Appl. Microbiol.* **7:** 58–66.

54 Zillig, W., I. Holz, D. Janekovic, H.P. Klenk, E. Imsel, J. Trent, S. Wunderl, V.H. Forjaz, R. Coutinho, and T. Ferreira. 1990. *Hyperthermus butylicus*, a hyperthermophilic sulfur-reducing archaebacterium that ferments peptides. *J. Bacteriol.* **172:** 3959–3965.

55 Zillig, W., I. Holz, and S. Wunderl. 1991. *Hyperthermus butylicus* gen. nov., sp. nov., a hyperthermophilic, anaerobic, peptide-fermenting, facultatively H_2S-generating archaeobacterium. *Int. J. Syst. Bacteriol.* **41:** 169–170.

56 De Rosa, M., A. Gambacorta, and J.D. Bu'Lock. 1975. Extremely thermophilic acidophilic bacteria convergent with *Sulfolobus acidocaldarius*. *J. Gen. Microbiol.* **86:** 156–164.

57 Nicolaus, B., A. Gambacorta, A.L. Basso, R. Riccio, and M. De Rosa, 1988. Trehalose in archaebacteria. *Syst. Appl. Microbiol.* **10:** 215–217.

58 Drioli, E., G. Iorio, G. Catapano, M. De Rosa, and A. Gambacorta. 1986. Capillary membrane reactors: Performances and applications. *J. Membr. Sci.* **27:** 253–361.

59 Nicolaus, B., A. De Simione, L. Del Piano, P. Giardina, and L. Lama. 1986. Production of 2-keto-3-deoxygluconate by immobilized cells of *Sulfolobus solfataricus*. *Biotechnol. Lett.* **8:** 497–500.

60 Segerer, A. and K.O. Stetter. 1991. The order *Sulfolobales*. In *The prokaryotes: A handbook on the biology of bacteria: ecophysiology, isolation, identification, applications*, 2nd ed. (ed. A. Balows et al.), pp. 684–701. Springer-Verlag, New York.

61 Lama, L., B. Nicolaus, A. Trincone, P. Morzillo, V. Calandrelli, and A. Gambacorta. 1991. Thermostable amylolytic activity from *Sulfolobus solfataricus*. *Biotechnol. Forum Europe* **8:** 201–203.

62 Pley, U., J. Seger, C.R. Woese, A. Gambacorta, H.W. Jannasch, H. Fricke, R. Rachel, and K.O. Stetter. 1991. *Pyrodictium abyssi* sp. nov.

represents a novel heterotrophic marine archaeal hyperthermophile growing at 110°C. *Syst. Appl. Microbiol.* **14:** 245–253.

63 Segerer, A.H., A. Trincone, M. Gahrtz, and K.O. Stetter. 1991. *Stygiolobus azoricus* gen. nov., sp. nov. represents a novel genus of anaerobic, extremely thermoacidophilic archaebacteria of the order *Sulfolobales. Int. J. Syst. Bacteriol.* **41:** 495–501.

64 Burggraf, S., H.W. Jannasch, B. Nicolaus, and K.O. Stetter. 1990. *Archaeoglobus profundus* sp. nov., represents a new species within the sulfate-reducing archaeobacteria. *Syst. Appl. Microbiol.* **13:** 24–28.

65 Huber, G. and K.O. Stetter. 1991. *Sulfolobus metallicus,* sp. nov., a novel strictly chemolithotrophic thermophilic archaeal species of metal-mobilizers. *Syst. Appl. Microbiol.* **14:** 372–378.

66 Validation of the publication of new names and new combinations previously effectively published outside the IJSB. List No. 40. 1992. *Int. J. Syst. Bacteriol.* **42:** 191–192.

67 Mathur, E., D. Shoemaker, B. Scott, J. Rombouts, M. Bergseid, and K. Nielson. 1992. *Pfu* DNA polymerase up date. Strategies in Molec. Biol. (News Letters, STRATAGENE) **5:** 11–13.

68 Völkl, P., R. Huber, E. Drobner, R. Rachel, S. Burggraf, A. Trincone, and K.O. Stetter. 1993. *Pyrobaculum aerophilum* sp. nov., a novel nitrate-reducing hyperthermophilic archaeum. *Appl. Environ. Microbiol.* **59:** 2918–2926.

69 Belkin, S. and H.W. Jannasch. 1985. A new extremely thermophilic, sulfur-reducing heterotrophic marine bacterium. *Arch. Microbiol.* **141:** 181–186.

70 Miroshnichenko, M.L., E.A. Bonch-Osmolovskaya, A. Neuner, N.A. Kostrikina, N.A. Chernych, and V.A. Alekseev. 1989. *Thermococcus stetteri* sp. nov., a new extremely thermophilic marine sulfur-metabolizing archaebacterium. *Syst. Appl. Microbiol.* **12:** 257–262.

71 Kobayashi, K., Y.S. Kwak, T. Akiba, T. Kudo, and K. Horikoshi. 1994. *Thermococcus profundus* sp. nov., a new hyperthermophilic archaeon isolated from a deep-sea hydrothermal vent. *Syst. Appl. Microbiol.* **17:** 232–236.

72 Bonch-Osmolovskaya, E.A., M.L. Miroshnichenko, N.A. Kostrikina, N.A. Chernych, and G.A. Zavarzin. 1990. *Thermoproteus uzoniensis* sp. nov., a new extremely thermophilic archaebacterium from Kmachatka continental hot springs. *Arch. Microbiol.* **154:** 556–559.

73 Neuner, A., H.W. Jannasch, S. Belkin, and K.O. Stetter. 1990. *Thermococcus litoralis* sp. nov.: A new species of extremely thermophilic marine archaebacteria. *Arch. Microbiol.* **153:** 205–207.

74 Validation of the publication of new names and combinations previously effectively published outside the IJSB. List No. 53. 1995. *Int. J. Syst. Bacteriol.* **45:** 418.

75 Klages, K.U. and H.W. Morgan. 1994. Characterisation of an extremely thermophilic sulphur-metabolising archaebacterium belonging to the Thermococcales. *Arch. Microbiol.* **162:** 261–266.

76 DeRosa, M., A. Gambacorta, B. Nicolaus, P. Giardina, E. Poerio, and V. Buonocore. 1984. Glucose metabolism in the extreme thermoacidophilic archaebacterium *Solfolobus solfataricus. Biochem. J.* **224:** 407–414.

Appendix 2

Media for Thermophiles

Compiled by F.T. Robb and A.R. Place

1. Growth Medium for Heterotrophic Hyperthermophilic Archaea

Artificial seawater (ASW) in g/liter distilled H_2O:

NaCl	20 g
$MgCl_2 \cdot 6H_2O$	3.0 g
$MgSO_4 \cdot 7H_2O$	6.0 g
$(NH_4)_2SO_4$	1.0 g
$NaHCO_3$	0.2 g
$CaCl_2 \cdot 2H_2O$	0.3 g
KCl	0.5 g
KH_2PO_4	0.42 g
NaBr	0.05 g
$SrCl_2 \cdot 6H_2O$	0.02 g
$Fe(NH_4)$ citrate	0.01 g

Added are 5.0 ml of Wolfe's trace minerals (see below) and 5.0 ml of vitamin mixture (see below). Some overlaps within the ASW and trace minerals ingredients are inconsequential.

Wolfe's trace minerals, stock solution (Wolin et al. 1963), in g/liter distilled H_2O:

Nitrilotriacetic acid (J.T. Baker R781-05)	1.5 g
$MgSO_4 \cdot 7H_2O$	3.0 g
$MnSO_4 \cdot 2H_2O$	0.5 g
NaCl	1.0 g
$FeSO_4 \cdot 7H_2O$	0.1 g
$CoCl_2$	0.1 g
$CaCl_2 \cdot 2H_2O$	0.1 g
$ZnSO_4$	0.1 g
$CuSO_4 \cdot 5H_2O$	0.01 g
$AlK(SO_4)_2$	0.01 g
H_3BO_3	0.01 g
$NaMoO_4 \cdot 2H_2O$	0.01 g

Vitamin mixture, stock solution (Bazylinski et al. 1989), in mg/liter distilled H_2O:

Niacin	10 mg
Biotin	4.0 mg
Pantothenate	10 mg
Lipoic acid	10 mg
Folic acid	4.0 mg
p-Aminobenzoic acid	10 mg
Thiamine (B_1)	10 mg

Riboflavin (B_2)	10 mg
Pyridoxine (B_6)	10 mg
Cobalamin (B_{12})	10 mg

Modified TYEG medium (Zeikus et al. 1979; Patel et al. 1985):

This medium consists of ASW (see above) containing 3 g/liter each of trypticase, glucose, and yeast extract. It contains no elemental sulfur and is sterilized by filtration, made anoxic by purging with oxygen-free (copper-purified) nitrogen (20 minutes), reduced by the addition of 3.2 mM $Na_2S \cdot 9H_2O$, and set at pH 7.2 with sterile HCl.

2. Enrichment Media

All reagents and solutions are made with Milli-Q deionized water (Millipore) or equivalent)

Synthetic seawater (per liter) (modified from "SME" medium of Stetter 1989; Pledger and Baross 1989)

NaCl	19.6 g
Na_2SO_4	3.3 g
KCl	0.5 g
KBr	0.05 g
H_3BO_3	0.02 g
$MgCl_2 \cdot 6H_2O$	8.8 g

Sulfolobus medium (per liter) (Brock et al. 1972) (see DSM 88 Medium)

Trace elements solutions (per liter)

Solution A (Pledger and Baross 1989)	
$CuSO_4 \cdot 5H_2O$	0.01 g
$ZnSO_4 \cdot 7H_2O$	0.1 g
$CoCl_2 \cdot 6H_2O$	0.005 g
$MnCl_2 \cdot 4H_2O$	0.2 g
$Na_2MoO_4 \cdot 2H_2O$	0.1 g
KBr	0.05 g
KI	0.05 g
H_3BO_3	0.1 g
LiCl	0.05 g
$Al_2(SO_4)_3$	0.05 g
$NiCl_2 \cdot 6H_2O$	0.01 g
Solution B (Pledger and Baross 1989)	
$VOSO_4 \cdot 2H_2O$	0.05 g
H_2WO_4	0.05 g
Na_2SeO_4	0.05 g
$NiCl_2 \cdot 6H_2O$	0.05 g
$SrCl \cdot 6H_2O$	0.05 g
$BaCl_2$	0.05 g

Vitamin solution (per liter) (Balch et al. 1979)

p-Aminobenzoic acid	5 mg
Biotin	2 mg
DL-Calcium pantothenate	5 mg
Cyanocobalamine (vitamin B_{12})	0.1 mg

Folic acid	2 mg
Nicotinic acid	5 mg
Pyridoxine-HCl	10 mg
Riboflavin	5 mg
Thiamine-HCl	5 mg
Lipoic acid	5 mg

Store in the dark at 2–4°C.

Methanothermus mineral solution 1 (per liter) (Balch et al. 1979)

K_2HPO_4	6 g

Methanothermus mineral solution 2 (per liter) (Balch et al. 1979)

KH_2PO_4	6 g
$(NH_4)_2SO_4$	6 g
NaCl	12 g
$MgSO_4 \cdot 7H_2O$	2.6 g
$CaCl_2 \cdot 2H_2O$	0.16 g

3. Plate Cultivation of Thermophiles

Double-strength artificial seawater (ASW; see Medium 1) medium (1 liter) as formulated in Protocol 1, except omitting $MgCl_2 \cdot 6H_2O$ and reducing $MgSO_4 \cdot 7H_2O$ to 1 g/liter

- Trace elements solution (20 ml/liter) (Balch et al. 1979) (see DSM 141 Medium)
- Yeast extract (2 g/liter)
- Peptone or Tryptone (8 g/liter)
- 2 ml Resazurin (0.1% solution)

Reducing agent

- $Na_2S \cdot 9H_2O$ (5%, w/v). Adjust pH to 7.5, filter-sterilize, and degas under oxygen-free N_2.

Culture and dilution medium

- Reduce medium with 5 ml of reducing agent per liter.
- For liquid culture, add 5 g of sulfur S^o (steam-sterilized 1 hour, three successive days) per liter of medium.

Polysulfides solution

- Dissolve 10 g of $Na_2S \cdot 9H_2O$ into 15 ml of deionized H_2O and add 3 g of sulfur flowers. Sterilize by filtration using a 0.2-μm pool size filter and keep under N_2.

4. Basal Salts Medium for *Pyrococcus furiosus*

The artificial seawater (ASW) was modified from a formulation described by Kester et al. (1967) and was used as a basis for several media formulations for thermoanaerobes.

Basic salts solution (in salt g/liter):

NaCl	15.0 g
$MgCl_2 \cdot 6H_2O$	1.0 g
Na_2SO_4	1.0 g
$CaCl_2 \cdot 2H_2O$	0.15 g
KCl	0.35 g
K_2HPO_4	0.14 g
NaBr	0.05 g
H_3BO_3	0.02 g
KI	0.02 g
$SrCl_2 \cdot 6H_2O$	0.01 g

Trace elements solution (in salt g/liter):

Nitrilotriacetic acid (J.T. Baker R781-05)	1.5 g
$MnSO_4 \cdot H_2O$	0.5 g
$FeSO_4 \cdot 7H_2O$	1.4 g
$NiCl_2 \cdot 6H_2O$	0.2 g
$CoSO_4$	0.1 g
$ZnSO_4 \cdot 7H_2O$	0.1 g
$CuSO_4 \cdot 5H_2O$	0.01 g
$Na_2MoO_4 \cdot 2H_2O$	0.01 g
$Na_2WO_4 \cdot 2H_2O$	0.3 g

Vitamin solution (vitamin mg/liter):

Biotin	2.0 mg
Folic acid	2.0 mg
Pyridoxine-HCl	10.0 mg
Thiamine-HCl	5.0 mg
Riboflavin	5.0 mg
Nicotinic acid	5.0 mg
DL-Ca-pantothenate	5.0 mg
Vitamin B12	0.1 mg
p-Aminobenzoic acid	5.0 mg
Lipoic acid	5.0 mg

The trace elements and vitamin solutions were added at 10 ml/liter when used. The trace elements solution may be added before autoclaving, and the vitamin solution must be filter-sterilized and added after autoclaving.

5. Growth Medium for *Methanococcus jannaschii*

The artificial seawater (ASW) for high-pressure cultivation of *Methanococcus jannaschii* was modified from a medium described by Jones et al. (1983) and contained the following components (per liter):

K_2HPO_4	0.14 g
$CaCl_2 \cdot 2H_2O$	0.14 g
NH_4Cl	0.25 g
$MgSO_4 \cdot 7H_2O$	3.4 g
$MgCl_2 \cdot 6H_2O$ ($\cdot 2H_2O$)	4.2 (2.7) g
KCl	0.33 g
NaCl	30.0 g
PIPES	15.12 g
1 mM SeO_2	1 ml
Minerals solution (see below)	10 ml

Adjust pH to 6.8 with NaOH. The trace minerals solution was modified from that described by Zeikus (1977) and contained the following components (per liter):

$FeCl_2 \cdot 4H_2O$	0.4 g
$MnCl_2 \cdot 4H_2O$	0.1 g
$CoCl_2 \cdot 6H_2O$	0.17 g
$ZnCl_2$	0.1 g
$CaCl_2 \cdot 2H_2O$	0.027 g
Nitrilotriacetic acid (J.T. Baker R781-05)	4.5 g
$NaMoO_4 \cdot 2H_2O$	0.01 g
$NiCl_2 \cdot 6H_2O$	0.036 g

Adjust pH to 7.0 with KOH. Prepare the ASW in advance, flush with N_2, seal under N_2 in anaerobic sample vials (100 ml per vial) (Balch and Wolfe 1976), and autoclave for sterility. Prior to inoculation, add a separate sulfur source (400 mM $Na_2S_2O_3 \cdot 5H_2O$, 10 ml/liter) and reducing agent (500 mM β-mercaptoethanol, 10 ml/liter). Substrate gas consisted of a 4:1 molar ratio of H_2 and CO_2, respectively.

6. Large-scale Growth Medium for *Pyrococcus furiosis*

The standard medium for *Pyrococcus furiosus* contains the following at concentrations given in g/liter unless indicated otherwise. The components are listed in the order of addition. The final pH is 6.8 (adjusted with KOH).

Maltose	5.0 g
NaCl	13.8 g
$MgSO_4$	3.5 g
$MgCl_2$	2.75 g
KCl	0.325 g
$CaCl_2$	0.75 g
KH_2PO_4	0.5 g
NaBr	0.05 g
H_3BO_4	0.015 g
$SrCl_2$	0.0075 g
Citric acid	0.005 g
KI	0.05 g
Resazurin	0.0025 g
Mineral medium (see below)	10 ml
Tryptone (Difco 0123-17-3)	5.0 g
Yeast Extract (Difco 0127-17-9)	5.0 g

Mineral medium contains (in g/liter unless indicated otherwise):

Nitrilotriacetic acid (J.T. Baker R781-05)	1.0 g
$MnSO_4$	0.5 g
$FeCl_3 \cdot 6H_20$	1.1 g
$Na_2WO_4 \cdot 2H_20$	0.3 g
$NiCl_2 \cdot 6H_20$	0.2 g
$CoSO_4 \cdot 7H_20$	0.1 g
$ZnSO_4 \cdot 7H_20$	0.1 g
$CuSO_4 \cdot 5H_20$	0.01 g
$Na_2MoO_4 \cdot 2H_20$	0.01 g
EDTA (J.T. Baker L704-05)	1.0 mM

7. Growth Medium for *Thermoplasma acidophilum* (Modified from Darland et al. 1970)

Medium 7A

Yeast extract (Difco 0127-17-9)	
or	
Yeast extract technical grade (Difco 0886-17-0)	1 g
Dextrose	10 g
KH_2PO_4	3 g
$MgSO_4$	0.5 g
$(NH_4)_2SO_4$	6.8 g
$CaCl_2 \cdot 2\ H_2O$	0.25 g

Adjust pH to 1.65 with H_2SO_4 (50%, v/v) (measured at room temperature). Sterilize the medium by heating to 100°C for 30 minutes (e.g., in a boiling water bath). Store at room temperature.

Medium 7B

*Yeast extract, technical grade (Difco 0886-17-0)	1 g
Sucrose	17 g
$MgSO_4$	0.5 g
$(NH_4)_2SO_4$	6.8 g
$CaCl_2 \cdot 2H_2O$	0.25 g
H_3PO_4 (85%)	1.5 ml
KOH (technical grade, 85% w/w)	1.22 g

*Reserve additional yeast extract (1 g/liter), to be added later.
Adjust pH to 1.60 with H_2SO_4 (50% v/v) (requires about 3 ml/liter medium). After 24 hours, add approximately 10 µl/liter of Antifoam A concentrate (Sigma A 5633).

8. Liquid Growth Medium for *Sulfolobus* (per liter)

Tryptone	2 g
or	
Sucrose	2 g
plus Yeast extract	1 g
$(NH_4)_2SO_4$	1.3 g
KH_2PO_4	0.28 g
$MgSO_4 \cdot 7H_2O$	0.25 g
$CaCl_2 \cdot 2H_2O$	0.07 g
$FeSO_4 \cdot 7H_2O$	0.028 g
$MnCl_2 \cdot 4H_2O$	1.8 mg
$Na_2B_4O_7 \cdot 10H_2O$	4.5 mg
$ZnSO_4 \cdot 7H_2O$	0.22 mg
$CuCl_2 \cdot 2H_2O$	0.05 mg
$NaMoO_4 \cdot 2H_2O$	0.03 mg
$VOSO_4 \cdot 2H_2O$	0.03 mg
$CoSO_4 \cdot 7H_2O$	0.01 mg

Adjust pH to 3–3.5 with H_2SO_4. Stock solutions are prepared separately for the different carbon sources, for $CaCl_2$, and for KH_2PO_4 plus trace elements. Stock solutions are prepared separately for the different carbon sources, for the $CaCl_2$, and for KH_2PO_4 plus trace elements. Medium for plates is solidified by gellan gum, 6.5 g per liter (Kelco 58-46-3070), and an additional 2 g of $MgCl_2 \cdot 6H_2O$ and 440 mg of $CaCl_2 \cdot H_2O$ (Grogan 1989).

9. Simplified Basal Medium for *Sulfolobus acidocaldarius*

2x basal medium (per liter):

K_2SO_4	6.0 g
NaH_2PO_4	1.0 g
$MgSO_4 \cdot 7H_2O$	0.6 g
$CaCl_2 \cdot 2H_2O$	0.2 g
Trace minerals solution (see below)	0.04 ml

Adjust pH to approximately 3.5 with H_2SO_4.

Trace minerals solution:

$FeCl_3 \cdot 6H_2O$	5.0 g
$CuCl_2 \cdot 2H_2O$	0.5 g
$CoCl_2 \cdot 6H_2O$	0.5 g
$MnCl_2 \cdot 4H_2O$	0.5 g
$ZnCl_2$	0.5 g

Dissolve in 100 ml of 1 N HCl.

10. Medium for cultivating ES4 (*Pyrococcus endeavori*) (Brown and Kelly 1989)

Solution A (per liter)

NaCl	47.8 g
Na_2SO_4	8.0 g
KCl	1.4 g
$NaHCO_3$	0.4 g
KBr	0.2 g
H_3BO_3	0.06 g

Solution B (per liter)

$MgCl_2 \cdot 6H_2O$	21.6 g
$CaCl_2 \cdot 2H_2O$	3.0 g
$SrCl_2 \cdot 6H_2O$	0.05g

Solution C (per liter)

NH_4CL	12.5 g
K_2HPO_4	7.0 g
CH_3CO_2Na	50.0 g

Individual solutions were prepared separately and sterilized for storage.

DSM 88. *Sulfolobus* Medium

$(NH_4)_2SO_4$	1.3 g
KH_2PO_4	0.28 g
$MgSO_4 \cdot 7H_2O$	0.25 g
$CaCl_2 \cdot 2H_2O$	0.07 g
$FeCl_3 \cdot 6H_2O$	0.02 g
$MnCl_2 \cdot 4H_2O$	1.8 mg
$Na_2B_4O_7 \cdot 10H_2O$	4.5 mg
$ZnSO_4 \cdot 7H_2O$	0.22 mg
$CuCl_2 \cdot 2H_2O$	0.05 mg
$Na_2MoO_4 \cdot 2H_2O$	0.03 mg
$VOSO_4 \cdot 2H_2O$	0.03 mg
$CoSO_4$	0.01 mg
Yeast extract (Difco 0127-17-9)	1.0 g
Freshly distilled H_2O	1000.0 ml

Adjust pH to 2.0 with 10 N H_2SO_4 at room temperature.

DSM 141. *Methanogenium* Medium

KCl	0.335 g
$MgCl_2 \cdot 6H_2O$	4.0 g
$MgSO_4 \cdot 7H_2O$	3.45 g

NH_4Cl	0.25 g
$CaCl_2 \cdot 2H_2O$	0.14 g
K_2HPO_4	0.14 g
NaCl	18.0 g
Trace elements (see below)	10.0 ml
Vitamin solution (see below)	10.0 ml
$Fe(NH_4)_2(SO_4)_2 \cdot 7H_2O$	2.0 mg
$NaHCO_3$	5.0 g
Na-acetate	1.0 g
Yeast extract (Difco 0127-17-9)	2.0 g
BBL-Trypticase (Fisher B11921)	2.0 g
Resazurin	1.0 mg
Cysteine hydrochloride	0.5 g
$Na_2S \cdot 9H_2O$	0.5 g
Distilled H_2O	1000.0 ml

Prepare the medium anaerobically under 80% H_2 + 20% CO_2 atmosphere. For incubation, use the same gas mixture at two atmospheres of pressure. If the medium is being used without gas mixture overpressure, then adjust pH with a little hydrochloric acid. Final pH should be 7.0 for strains *DSM 1497, DSM 1537, DSM 2067, DSM 2279,* and *DSM 2373*; 6.8 for strain *DSM 2095,* and 6.5 for strain *DSM 1498.* For *DSM 2373,* increase the amount of trypticase to 6 g/liter.

Trace elements solution:

Nitrilotriacetic acid	1.5 g
$MgSO_4 \cdot 7H_2O$	3.0 g
$MnSO_4 \cdot 2H_2O$	0.5 g
NaCl	1.0 g
$FeSO_4 \cdot 7H_2O$	0.1 g
$CoSO_4 \cdot 7H_2O$	0.18 g
$CaCl_2 \cdot 2H_2O$	0.1 g
$ZnSO_4 \cdot 7H_2O$	0.18 g
$CuSO_4 \cdot 5H_2O$	0.01 g
$KAl(SO_4)_2 \cdot 12H_2O$	0.02 g
H_3BO_3	0.01 g
$Na_2MoO_4 \cdot 2H_2O$	0.01 g
$NiCl_2 \cdot 6H_2O$	0.025 g
$Na_2SeO_3 \cdot 5H_2O$	0.3 mg
Distilled H_2O	1000.0 ml

First dissolve nitrilotriacetic acid and adjust pH to 6.5 with KOH, then add minerals. Final pH 7.0 (with KOH).

Vitamin solution:

Biotin	2.0 mg
Folic acid	2.0 mg
Pyridoxine-HCl	10.0 mg
Thiamine-HCl	5.0 mg
Riboflavin	5.0 mg
Nicotinic acid	5.0 mg
DL-Calcium pantothenate	5.0 mg
Vitamin B_{12}	0.1 mg
p-Aminobenzoic acid	5.0 mg
Lipoic acid	5.0 mg
Distilled H_2O	1000.0 ml

DSM 150. *Acidianus brierleyi* Medium

$(NH_4)_2SO_4$	3.0 g
$K_2HPO_4 \cdot 3H_2O$	0.5 g
$MgSO_4 \cdot 7H_2O$	0.5 g
KCl	0.1 g
$Ca(NO_3)_2$	0.01 g
Yeast extract (Difco 0127-17-9)	0.2 g
Sulfur (flowers)	10.0 g
Distilled H_2O	1000.0 ml

Adjust pH to 1.5–2.5 with 6 N H_2SO_4. Autoclave yeast extract (10% [w/v] in distilled H_2O) separately. Sterilize sulfur by steaming for 3 hours on each of three successive days.

DSM 158. *Thermoplasma acidophilum* medium

$(NH_4)_2SO_4$	1.32 g
KH_2PO_4	0.372 g
$MgSO_4 \cdot 7H_2O$	0.247 g
$CaCl_2 \cdot 2H_2O$	0.074 g
Trace elements solution (see below)	10.0 ml
Yeast extract (Difco 0127-17-9)	1.0 g
Glucose	10.0 g
Freshly distilled H_2O	1000.0 ml

Adjust pH to 1.0–2.0 with 10 N H_2SO_4. Sterilize separately stock solutions of yeast extract (10% w/v) and glucose (50% w/v) in H_2O by autoclaving and thereafter add to the sterile mineral salts medium.

Trace elements solution:

$FeCl_3 \cdot 6H_2O$	1.93 g
$MnCl_2 \cdot 4H_2O$	0.18 g
$Na_2B_4O_7 \cdot 10H_2O$	0.45 g
$ZnSO_4 \cdot 7H_2O$	22.0 mg
$CuCl_2 \cdot 2H_2O$	5.0 mg
$Na_2MoO_4 \cdot 2H_2O$	3.0 mg
$VOSO_4 \cdot 5H_2O$	3.8 mg
$CoSO_4 \cdot 7H_2O$	2.0 mg
Distilled H_2O	1000.0 ml

DSM 182. *Sulfolobus solfataricus* Medium

Yeast extract (Difco 0127-17-9)	1.0 g
Casamino acids (Difco 0230-01-1)	1.0 g
KH_2PO_4	3.1 g
$(NH_4)_2SO_4$	2.5 g
$MgSO_4 \cdot 7H_2O$	0.2 g
$CaCl_2 \cdot 2H_2O$	0.25 g
$MnCl_2 \cdot 4H_2O$	1.8 mg
$Na_2B_4O_7 \cdot 10H_2O$	4.5 mg
$ZnSO_4 \cdot 7H_2O$	0.22 mg
$CuCl_2 \cdot 2H_2O$	0.05 mg
$Na_2MoO_4 \cdot 2H_2O$	0.03 mg
$VOSO_4 \cdot 2H_2O$	0.03 mg
$CoSO_4 \cdot 7H_2O$	0.01 mg
Distilled H_2O	1000.0 ml

Adjust pH to 4.0–4.2 with 10 N H_2SO_4 at room temperature.

DSM 184. ***Desulfurococcus*** **Medium**

To DSM Medium 88, add (per liter):

Yeast extract (Difco 0127-17-9)	1.0 g
Resazurin	1.0 mg
Sulfur (powdered)	5.0 g
$Na_2S \cdot 9H_2O$	0.5 g

Adjust pH to 5.5 with 10 N H_2SO_4. Sterilize separately Na_2S (autoclave stock solution under N_2) and sulfur (steaming for 3 hours each of three successive days). Prepare the medium anaerobically (100% nitrogen).

DSM 185. ***Thermoproteus*** **Medium**

$(NH_4)_2SO_4$	0.264 g
$FeSO_4 \cdot 7H_2O$	0.556 g
$MgSO_4 \cdot 7H_2O$	0.492 g
$CaSO_4 \cdot 2H_2O$	0.344 g
KH_2PO_4	0.014 g
Resazurin	1.0 mg
Trace elements (amounts taken from 0.1% [w/v] solutions):	
NaF	0.84 ml
$MnCl_2 \cdot 4H_2O$	0.18 ml
$Na_2B_4O_7 \cdot 10H_2O$	0.45 ml
$ZnSO_4 \cdot 7H_2O$	0.022 ml
$CuCl_2 \cdot 2H_2O$	0.005 ml
$Na_2MoO_4 \cdot 2H_2O$	0.003 ml
$CoSO_4 \cdot 7H_2O$	0.001 ml
Yeast extract (Difco 0127-17-9)	0.2 g
Soluble starch	5.0 g
Sulfur (powered)	10.0 g
$Na_2S \cdot 9H_2O$	0.5 g
Distilled H_2O	1000.0 ml

Final pH is 5.5. Gas atmosphere is 100% N_2. Prepare the medium without starch, Na_2S, and sulfur, adjust pH to 5.5, and filter-sterilize. Add 50 ml of 10% (w/v) autoclaved starch solution (dissolved by heating in boiling water bath) and flush with oxygen-free nitrogen gas for 30 minutes. Then add 10 ml of 5% (w/v) sterile Na_2S solution (separately autoclaved under nitrogen) and adjust pH of the medium to 5.5 with sterile 1 N H_2SO_4. Distribute the medium into tubes containing the appropriate amount of sterile sulfur powder (sterilized by steaming for 3 hours on each of three successive days).

DSM 193. ***Desulfobacter postgatei*** **Medium**

Solution A:

Na_2SO_4	3.0 g
KH_2PO_4	0.2 g
NH_4Cl	0.3 g
NaCl	7.0 g
$MgCl_2 \cdot 6H_2O$	1.3 g
KCl	0.5 g
$CaCl_2 \cdot 2H_2O$	0.15 g
Resazurin	1.0 mg
Distilled H_2O	870.0 ml

Solution B:

Trace elements solution SL-10 (see DSM Medium 320)	1.0 ml

Solution C:

$NaHCO_3$	5.0 g
Distilled H_2O	100.0 ml

Solution D:

Sodium acetate·$3H_2O$	2.5 g
Distilled H_2O	10.0 ml

Solution E:

Vitamin solution (see Medium 2)	10.0 ml

Solution F:

$Na_2S \cdot 9H_2O$	0.4 g
Distilled H_2O	10.0 ml

Boil Solution A for a few minutes, cool to room temperature, gas with 80% N_2 + 20% CO_2 gas mixture to reach a pH below 6 and then autoclave anaerobically under the same gas mixture. Autoclave Solutions B, D, E, and F separately under nitrogen. Filter-sterilize Solution C and flush with 80% N_2 + 20% CO_2 to remove dissolved oxygen. Add Solutions B through F to the sterile, cooled Solution A in the sequence as indicated. The complete medium is distributed anaerobically under 80% N_2 + 20% CO_2 into appropriate vessels. Final pH of the medium is 7.1–7.4. Addition of 10–20 mg of sodium dithionite per liter (e.g., from 5% [w/v] solution, freshly prepared under N_2 and filter-sterilized) may stimulate growth at the beginning. For transfers, use 5–10% inoculum.

DSM 195. *Desulfobacter* sp. Medium

Use DSM Medium 193, but change the amount of NaCl to 21.0 g/liter and of $MgCl_2 \cdot 6H_2O$ to 3.1 g/liter. For strain *DSM 4661:* Replace sodium acetate by 1 mM resorcinol, added from a freshly prepared, filter-sterilized, anaerobic stock solution. During growth, feed the culture once with the same amount of resorcinol.

DSM 265. *Thermofilum pendens* Medium

$(NH_4)_2SO_4$	1.3 g
KH_2PO_4	0.28 g
MgSO4·$7H_2O$	0.25 g
$CaCl_2 \cdot 2H_2O$	0.07 g
$FeCl_3 \cdot 6H_2O$	0.02 g
$MnCl_2 \cdot 4H_2O$	1.8 mg
$Na_2B_4O_7 \cdot 10H_2O$	4.5 mg
$ZnSO_4 \cdot 7H_2O$	0.22 mg
$CuCl_2 \cdot 2H_2O$	0.05 mg
$Na_2MoO_4 \cdot 2H_2O$	0.03 mg
$VOSO_4 \cdot 2H_2O$	0.03 mg
$CoSO_4 \cdot 7H_2O$	0.01 g
Yeast extract (Difco 0127-17-9)	2.0 g
Sucrose	2.0 g
Sulfur (powdered)	10.0 g
Polar lipid fraction prepared from *Thermoproteus tenax* (DSM 2078) or from any other Archaea, aqueous suspension	6–12.0 ml
$Na_2S \cdot 9H_2O$	0.3 g
Distilled H_2O	1000.0 ml

Adjust final pH to 5.2. Prepare the medium anaerobically under 100% nitrogen. Prepare the following constituents separately and add to the autoclaved mineral salts solution: Yeast extract (20 ml of 10% [w/v] solution), boiled for few minutes, not autoclaved; sucrose (20 ml of 10% [w/v] solution), filter-sterilized; sulfur (10 g), sterilized by steaming for 3 hours on each of three successive days; polar lipid fraction (6–12 ml), prepared as described by W. Zillig et al. (*Syst. Appl. Microbiol. 4:* 79–87 [1983]); $Na_2S{\cdot}9H_2O$ (10 ml of 3% [w/v] solution), autoclaved under nitrogen atmosphere.

DSM 266. *Thermococcus celer* Medium

$(NH_4)_2SO_4$	1.3 g
KH_2PO_4	0.28 g
$MgSO_4{\cdot}7H_2O$	0.25 g
$CaCl_2{\cdot}2H_2O$	0.07 g
$FeCl_3{\cdot}6H_2O$	0.02 g
$MnCl_2{\cdot}4H_2O$	1.8 mg
$Na_2B_4O_7{\cdot}10H_2O$	4.5 mg
$ZnSO_4{\cdot}7H_2O$	0.22 mg
$CuCl_2{\cdot}2H_2O$	0.05 mg
$Na_2MoO_4{\cdot}2H_2O$	0.03 mg
$VOSO_4{\cdot}2H_2O$	0.03 mg
$CoSO_4{\cdot}7H_2O$	0.01 mg
NaCl	40.0 g
Resazurin	1.0 mg
Yeast extract (Difco 0127-17-9)	2.0 g
Sulfur (powdered)	5.0 g
Distilled H_2O	1000.0 ml

Adjust final pH to 5.8. Prepare the medium anaerobically under 100% nitrogen. Prepare the following constituents separately and add to the autoclaved mineral salts solution: Yeast extract (20 ml of 10% [w/v] solution), boiled for a few minutes not autoclaved; sulfur (10 g), sterilized by steaming for 3 hours on each of three successive days; $Na_2S{\cdot}9H_2O$ (10 ml of 3% [w/v] solution), autoclaved under nitrogen atmosphere.

DSM 283. *Pyrodictium* Medium

NaCl	13.85 g
$MgSO_4{\cdot}7H_2O$	3.5 g
$MgCl_2{\cdot}6H_2O$	2.75 g
KCl	0.325 g
NaBr	0.05 g
H_3BO_3	0.015 g
$SrCl_2{\cdot}6H_2O$	7.5 mg
$(NH_4)_2SO_4$	10.0 mg
Citric acid	5.0 mg
KI	0.05 mg
$CaCl_2{\cdot}2H_2O$	0.75 g
KH_2PO_4	0.5 g
$NiCl_2{\cdot}6H_2O$	2.0 mg
Trace minerals solution (see DSM Medium 141)	10.0 ml
Resazurin	1.0 mg
Yeast extract (Difco 0127-17-9):	
for *P. occultum*	0.2 g
for *P. brockii*	2.0 g

Sulfur (powdered)	30.0 g
$Na_2S \cdot 9H_2O$	0.5 g
Distilled H_2O	1000.0 ml

Adjust pH to 5.5 with 10 N sulfuric acid. Prepare the medium without Na_2S, boil, and then cool down under a stream of 80% H_2 + 20% CO_2. Add Na_2S, adjust the pH to 5.5, and distribute in appropriate vessels under H_2/CO_2, thereby taking care to transfer also the necessary amount of sulfur. Heat the vessels containing the medium in boiling water for 1 hour before inoculation. For storage of medium at room temperature, heat vessels at least on two successive days. Do not autoclave! After inoculation, pressurize the vessels to 200–300 kPa H_2/CO_2 gas mixture.

Alternative *Pyrodictium* Medium

Seawater (natural or prepared from salt for seawater aquarium)	1000.0 ml
Sulfur (powdered)	30.0 g
Trace minerals solution (see DSM Medium 141)	10.0 ml
KH_2PO_4	0.5 g
$NiCl_2 \cdot 6H_2O$	2.0 mg
Yeast extract (Difco 0127-17-9):	
for *P. occultum*	0.2 g
for *P. brockii*	2.0 g
Resazurin	1.0 mg
$Na_2S \cdot 9H_2O$	0.5 g

Adjust pH to 5.5 before boiling. Readjust pH to 5.5 after Na_2S has been added. Proceed as indicated above.

DSM 291. *Thermoproteus neutrophilus* Medium

Prepare DSM Medium 88 without yeast extract, then add per liter:

Sulfur (powdered)	8.0 g
Resazurin	0.4 mg

Adjust pH to 6.5 with NaOH, bring to a boil for 5 minutes, and then cool to room temperature under H_2/CO_2 (80:20) gas mixture. Add 0.85 g of $NaHCO_3$ per liter and equilibrate the pH to about 6.5 by further gassing. Dispense in serum bottles (10 ml/30 ml bottle), seal, and heat the bottles for 1 hour at 85°C on each of three successive days. Before inoculation, add from anaerobic stock solutions 0.02% yeast extract (for mixotrophic growth only) and 0.01 ml of sodium dithionite solution (250 mg dithionite/10 ml of distilled H_2O, filter-sterilized). After inoculation, pressurize to 1 bar H_2/CO_2 overpressure.

DSM 348. SF1 Medium

$MgCl_2 \cdot 6H_2O$	7.0 g
$MgSO_4 \cdot 7H_2O$	6.0 g
$CaCl_2 \cdot 2H_2O$	0.5 g
$K_2HPO_4 \cdot 3H_2O$	0.4 g
NH_4Cl	1.0 g
NaCl	120.0 g
KCl	3.8 g
Trace elements solution SL-10 (see DSM Medium 320)	1.0 ml
$Na_2SeO_3 \cdot 5H_2O$	75.0 µg
Yeast extract (Difco 0127-17-9)	2.0 g

BBL-Trypticase (Fisher B11921)	2.0 g
Resazurin	1.0 mg
Cysteine hydrochloride	0.5 g
10 M NaOH	0.6 ml
Na_2CO_3	1.0 g
Trimethylamine hydrochloride	1.9 g
$Na_2S{\cdot}9H_2O$	0.25 g
Distilled H_2O	1000.0 ml

Final pH is 7.3–7.4. Gas atmosphere is 80% N_2 + 20% CO_2. Dissolve ingredients except cysteine, NaOH, Na_2CO_3 trimethylamine, and Na_2S in water, boil for 5 minutes, and cool to room temperature while gassing with 80% N_2 + 20% CO_2. Add cysteine and NaOH, equilibrate pH to 6.7, distribute into anaerobic tubes under the same gas, and autoclave. Thereafter, add per liter from anaerobic stock solutions autoclaved under nitrogen: Trimethylamine hydrochloride, 1.9 g in 20 ml; Na_2CO_3, 1.0 g in 20 ml; $Na_2S{\cdot}9H_2O$, 0.25 g in 10 ml of water. Control final pH.

DSM 358. *Acidianus infernus* Medium

Aerobic growth: Use DSM Medium 88 with 1.0 g/liter yeast extract and 1.0 g/liter sulfur and adjust pH to 2.5. Cultivate under air enriched with 1–10% CO_2.

Anaerobic growth: Use DSM Medium 88 with 0.5 g/liter yeast extract, 1.0 g/liter sulfur, 1.0 mg/liter resazurin and adjust pH to 2.5. Prepare the medium under 80% H_2 + 20% CO_2 gas mixture. Pressurize inoculated bottles to 100 kPa H_2/CO_2. Sterilize yeast extract and sulfur separately (steam sulfur for 3 hours on each of three successive days).

For strain *DSM 6334*: Prepare aerobic and anaerobic media with 5.0 g/liter sulfur and adjust pH to 2–2.5. Use 1 g/liter of yeast extract for aerobic medium and 0.2 g/liter for anaerobic medium. Pressurize inoculated bottles to 200 kPa H_2/CO_2 (80:20).

For strain *DSM 3772*: Use aerobic media with 0.02 g/liter and anaerobic media with 0.002 g/liter of yeast extract.

DSM 376. AN1 Medium

BBL-Trypticase (Fisher B11921)	10.0 g
K_2HPO_4	1.5 g
NaCl	2.5 g
Sodium thioglycolate	1.0 g
Sulfur (powdered)	8.0 g
Resazurin	1.0 mg
Distilled H_2O	1000.0 ml

Adjust pH to 7.3. Prepare the medium anaerobically under 100% nitrogen. Sterilize thioglycolate and sulfur separately. Sterilize the sulfur by steaming for 3 hours on each of three successive days.

DSM 377. *Pyrococcus/Staphylothermus* Medium

Use DSM Medium 283 supplemented with 1 g/liter yeast extract (Bacto, Difco 0127-17-9) and 5 g/liter peptone (Bacto, Difco 0123-17-3). Prepare the medium under 100% nitrogen gas. Adjust pH to 6.5.

Alternative Medium:

KH_2PO_4	0.5 g
$NiCl_2{\cdot}6H_2O$	2.0 mg
Trace elements solution (see DSM Medium 141)	10.0 ml
Sulfur (powdered)	30.0 g
Yeast extract (Difco 0127-17-9)	1.0 g
Bacto-peptone (Difco 0123-17-3)	5.0 g
Resazurin	1.0 mg
Seawater	1000.0 ml

Adjust pH to 6.3–6.5, boil, cool, and dispense under nitrogen. Before use, reduce the medium with a sterile neutral solution of $Na_2S{\cdot}9H_2O$ to 0.5 g/liter final concentration. Check that final pH is 6.5.

DSM 387. *Methanothrix* (Thermophilic) Medium

NH_4Cl	0.5 g
K_2HPO_4	0.4 g
$MgCl_2{\cdot}6H_2O$	0.1 g
Resazurin	1.0 mg
Trace elements solution (see DMS Medium 141)	10.0 ml
Distilled H_2O	1000.0 ml

Bring to a boil for 10 minutes and then cool to room temperature under 80% N_2 + 20% CO_2 gas mixture. Gas the medium until a pH of 5.8 is reached and then dispense in serum bottles under the same gas. Autoclave and then add from anaerobic sterile stock solutions per liter:

5% $NaHCO_3$ solution (outgassed with N_2/CO_2 for 15 min)	20.0 ml
1% $CaCl_2{\cdot}2H_2O$ solution	10.0 ml
33% sodium acetate solution	10.0 ml
Vitamin solution (see Medium 2)	10.0 ml
1.42% coenzyme M solution	10.0 ml
5% $Na_2S{\cdot}9H_2O$ solution	5.0 ml

Finally, add additional CO_2 gas (by syringe and injection) to bring the gas atmosphere to 30% CO_2. For instance, to 30-ml serum bottles containing 15 ml of medium and 15 ml gas phase of 80% N_2 + 20% CO_2, 1.5 ml of CO_2 is added. Final pH should be 6.5.

DSM 390. *Pyrobaculum* Medium

Use DSM Medium 88 without yeast extract, but add:

Resazurin	1.0 mg/liter
Bacto-peptone (Difco 0123-17-3; other peptones may be less suitable)	0.5 g/liter
Bacto-yeast extract (Bacto, Difco 0127-17-9)	0.2 g/liter
$Na_2S_2O_3{\cdot}5H_2O$ (for *DSM 4184* only)	2.0 g/liter
or	
Sulfur (powdered; for *DSM 4185* only)	20.0 g/liter
$Na_2S{\cdot}9H_2O$	0.5 g/liter

Dissolve ingredients (mineral salts, resazurin, sulfur compound) in water and adjust pH to 6.0 with 8 N NaOH. Gas the medium with N_2 for 30 minutes, add peptone, yeast extract, and Na_2S, readjust pH to 6.0 with 10 N H_2SO_4, and dispense under N_2. For storage, heat the medium to 90°C for 1 hour on each of three successive days. Do not autoclave the medium.

DSM 391. WMC Medium

Mineral salts solution "2xW" (see below)	500.0 ml
Trace elements solution (see DSM Medium 141)	10.0 ml
"LIP" solution (see below)	50.0 ml
Sodium acetate	1.0 g
Resazurin	1.0 mg
$NaHCO_3$	5.0 g
L-Cysteine	0.5 g
Distilled H_2O	450.0 ml

pH is 7.2. Prepare the medium anaerobically under 80% H_2 + 20% CO_2 gas mixture. Before use, add from sterile anaerobic stock solutions 1/20 volume of "TYC" (see below) and 0.5 g/liter of $Na_2S \cdot 9H_2O$.

For strain *DSM 4254:* Add a filter-sterilized, anaerobic solution of L-histidine to a final concentration of 80 mg/liter.

For strain *DSM 4310:* Add 32 mg/liter of hypoxanthine to the medium before autoclaving; after heat sterilization, add 140 μg/liter of vitamin B_{12} (cyanocobalamine) from an anaerobic filter-sterilized stock solution.

Mineral salts solution "2xW":

NaCl	40.0 g
$MgCl_2 \cdot 6H_2O$	5.6 g
$MgSO_4 \cdot 7H_2O$	0.7 g
KCl	0.68 g
NH_4Cl	0.5 g
K_2HPO_4	0.28 g
$CaCl_2 \cdot 2H_2O$	0.28 g
Distilled H_2O	1000.0 ml

May be stored at room temperature in the dark.

"LIP" solution:

L-Leucine	5.0 g
L-Isoleucine	10.0 g
Pantothenate	0.1 g
Distilled H_2O	1000.0 ml

May be stored frozen without sterilization.

"TYC" solution:

Casamino acids	100.0 g
Yeast extract (Difco 0127-17-9)	50.0 g
L-Tryptophan	1.0 g
Distilled H_2O	1000.0 ml

Prepare and autoclave anaerobically under N_2.

DSM 395. *Desulfurococcus amylolyticus* Medium

NH_4Cl	0.33 g
KH_2PO_4	0.33 g
KCl	0.33 g
$CaCl_2 \cdot 2H_2O$	0.44 g
$MgCl_2 \cdot 6H_2O$	0.7 g
NaCl	0.5 g
Trace elements solution SL-10 (see DSM Medium 320)	1.0 ml

Vitamin solution (see Medium 2)	10.0 ml
Yeast extract (Difco 0127-17-9)	0.2 g
Starch or peptone or casein hydrolysate (BBL-Trypticase Fisher B11921)	5.0 g
Sulfur (powdered)	10.0 g
Resazurin	1.0 mg
$NaHCO_3$	0.8 g
$Na_2S{\cdot}9H_2O$	0.5 g
Distilled H_2O	1000.0 ml

Adjust pH to 6.2–6.4. Prepare medium anaerobically under 80% N_2 + 20% CO_2 gas phase.

DSM 398. ***Thermoplasma volcanium* Medium**

KH_2PO_4	3.0 g
$MgSO_4{\cdot}7H_2O$	1.0 g
$CaCl_2{\cdot}2H_2O$	0.25 g
$(NH_4)_2SO_4$	0.2 g
Yeast extract (Difco 0127-17-9)	1.0 g
Glucose	5.0 g
Sulfur (for anaerobic media only)	4.0 g
Distilled H_2O	1000.0 ml

Adjust pH of the mineral base to about 2 with sulfuric acid and autoclave. Add glucose and yeast extract from filter-sterilized stock solutions.

For *DSM 4301:* Add in addition 0.5 g/liter meat extract (Merck); aerobic growth of that strain is sometimes difficult to achieve. Check that pH is 2–3.

For *anaerobic growth:* Distribute acidified mineral-base-containing sulfur in serum bottles (20 ml per 100-ml bottle) under N_2/CO_2 (80:20) gas mixture. Sterilize the bottles by tyndallization. Add yeast extract, glucose, and meat extract (for *DSM 4301* only) from separately sterilized stock solutions.

DSM 399. ***Archaeoglobus* Medium**

KCl	0.34 g
$MgCl_2{\cdot}6H_2O$	4.0 g
$MgSO_4{\cdot}7H_2O$	3.45 g
NH_4Cl	0.25 g
$CaCl_2{\cdot}2H_2O$	0.14 g
K_2HPO_4	0.14 g
NaCl	18.0 g
$NaHCO_3$	5.0 g
Yeast extract (Difco 0127-17-9)	0.5 g
Sodium-L-lactate	1.5 g
$Fe(NH_4)_2(SO_4)_2{\cdot}7H_2O$	2.0 mg
Trace elements solution (see DSM Medium 141)	10.0 ml
Resazurin	1.0 mg
$Na_2S{\cdot}9H_2O$	0.5 g
Distilled H_2O	1000.0 ml

Dissolve ingredients (except bicarbonate and sulfide), bring to a boil for few minutes, then cool quickly to room temperature while gassing with N_2/CO_2 (80:20), add sodium bicarbonate, and adjust pH to 6.9. Distribute in serum bottles (20 ml per 100 ml bottle) under N_2/CO_2, seal, pressurize bottles up to 2 bar overpressure, and then autoclave. Before use, reduce the medium with Na_2S from anaerobic sterile neutral stock solution.

DSM 480. ***Desulfurella* Medium**

NH_4Cl	0.33 g
$CaCl_2 \cdot 2H_2O$	0.33 g
$MgCl_2 \cdot 6H_2O$	0.33 g
KCl	0.33 g
KH_2PO_4	0.33 g
Trace elements solution SL-10 (see DSM Medium 320)	1.0 ml
Vitamin solution (see Medium 2)	10.0 ml
Yeast extract (Difco 0127-17-9)	0.1 g
Sodium acetate	5.0 g
Sulfur (powdered)	10.0 g
$NaHCO_3$	2.0 g
Resazurin	1.0 mg
$Na_2S \cdot 9H_2O$	0.5 g
Distilled H_2O	1000.0 ml

Prepare medium (without bicarbonate, vitamins, Na_2S) anaerobically under 80% N_2 + 20% CO_2. Adjust pH to 5.9 before sterilization. Sterilize medium by heating for 1 hour at 90–100°C on three subsequent days. Before use, add to the medium 40 ml/liter of 5% (w/v) sterile, anaerobic $NaHCO_3$ solution, vitamins, and Na_2S. Medium pH is 6.8–7.0.

DSM 480b. TTD Medium

Use DSM Medium 480 without yeast extract, sodium acetate, and $NaHCO_3$. Add to the medium 0.5% casitone and 2.5% NaCl (marine source). Adjust pH to 5.7 with hydrochloride acid before sterilization. Sterilize the medium by heating for 1 hour at 90–100°C on three subsequent days. Before use, add 30.0 ml/liter of 5% (w/v) sterile, anaerobic $NaHCO_3$ solution, vitamins, and sulfide. Final pH should be 6.5.

DSM 491. ***Hyperthermus butylicus* Medium**

Use DSM Medium 283 (prepared with synthetic seawater) with:

NaCl	17.0 g/liter
KI	2.5 mg/liter
NH_4Cl	0.5 g/liter
Sulfur (powdered)	6.0 g/liter
Tryptone (Difco 0123-17-3)	6.0 g/liter

Adjust pH to 6.0 with sulfuric acid. Prepare medium under 100% nitrogen gas. Sterilize medium by heating for 1 hour at 90°C on three subsequent days. Before use, reduce the medium by adding 0.3 g/liter of $Na_2S \cdot 9H_2O$ from a sterile anaerobic stock solution. Check that pH is 6.0–6.5. After inoculation, pressurize the culture vessel to 1 bar H_2/CO_2 (80:20) overpressure.

DSM 508. ***Pyrodictium abyssi* Medium**

Use DSM Medium 283 with 0.5 g/liter of yeast extract and 0.1 mg/liter of $Na_2WO_4 \cdot 2H_2O$. Omit citric acid. Use 0.75 g/liter of $CaCl_2 \cdot 2H_2O$ (not 1.0 mg as indicated in DSM catalog 1989). Medium pH is 5.5–6.0.

DSM 510. ***Stygiolobus*** **Medium**

Use anaerobic DSM Medium 358 with 0.2 g/liter of yeast extract, 5 g/liter of sulfur, 0.5 mg/liter of resazurin, and adjust pH to 2.5–3.0. Pressurize inoculated bottles (20 ml of medium per 100-ml serum bottle) to 200 kPa H_2/CO_2 (80:20).

DSM 519. ***Archaeoglobus profundus*** **Medium**

Use DSM Medium 399 with 1 g/liter of sodium acetate and 2.7 g/liter of sodium sulfate added and with only 1 g/liter of sodium bicarbonate. Omit the sodium lactate. Medium pH should be 6.5. Pressurize vessels to 2 bar H_2/CO_2 (80:20) overpressure.

DSM 525. MG-Medium

KCl	0.335 g
$MgCl_2 \cdot 6H_2O$	2.75 g
$MgSO_4 \cdot 7H_2O$	3.45 g
NH_4Cl	0.25 g
$K_2HPO_4 \cdot 3H_2O$	0.14 g
$CaCl_2 \cdot 2H_2O$	0.14 g
NaCl	100.0 g
Trace minerals solution (see DSM Medium 141)	10.0 ml
Sodium acetate	1.0 g
Resazurin	1.0 mg
Trimethylamine x HCl	5.0 g
$NaHCO_3$ (5% w/v)	80.0 ml
Na_2CO_3 (5% w/v)	10.0 ml
Vitamin solution (see Medium 2)	10.0 ml
Cysteine hydrochloride	0.5 g
$Na_2S \cdot 9H_2O$	0.5 g
Distilled H_2O	1000.0 ml

Adjust pH to 6.9–7.0. Dissolve ingredients except trimethylamine, carbonates, vitamins, cysteine hydrochloride, and Na_2S. Flush the medium first with N_2 for 20 minutes, then with 80% N_2 + 20% CO_2 for 10 minutes, dispense under N_2/CO_2, and autoclave. Sterilize separately stock solutions of trimethylamine x HCl (gassed and autoclaved under N_2), $NaHCO_3$ (gassed with 80% N_2 + 20% CO_2 and autoclaved), Na_2CO_3 (gassed with N_2 and autoclaved), vitamins (filter-sterilized and gassed with N_2), cysteine, and sulfide (each autoclaved under N_2).

For *DSM 5700, DSM 5701, DSM 5702, DSM 5703,* and *DSM 5814*: Increase the amount of NaCl to 150 g/liter.

DSM 560. MTP4 Medium

Use DSM Medium 195 without Na_2SO_4 and sodium acetate or DSM Medium 141 with N_2/CO_2 (80:20) gas phase. Substrate is methanol (20–30 mM) plus 1–2 ml of methanethiol gas per 50 ml of culture. Use 0.4 g/liter of Na_2S and dithionite (see DSM Medium 193) as reducing agents.

DSM 611. BS-Medium

Marine medium/Synthetic seawater mix (see below)	125.0 ml
Trace elements solution (see DSM Medium 141)	10.0 ml

Na_2SeO_4	0.1 mg
$Na_2WO_4 \cdot 2H_2O$	0.1 mg
$(NH_4)_2(Fe(SO_4)_2 \cdot 6H_2O$	0.2 mg
NH_4Cl	0.25 g
KH_2PO_4	0.07 g
$NaHCO_3$	2.2 g
Resazurin	0.5 mg
Distilled H_2O	1000.0 ml

Dissolve ingredients, adjust pH to 7.0 with H_2SO_4, and dispense in serum bottles (20 ml medium per 100-ml bottle). Replace gas phase by N_2/CO_2 (80:20) and autoclave. Before use, add yeast extract (5% stock solution) and KNO_3 (10% stock solution) to a final concentration of 0.05% and 0.1%, respectively. After inoculation, pressurize bottles up to 2 bar N_2/CO_2 or H_2/CO_2 (80:20) overpressure.

Marine medium/Synthetic seawater mix:

NaCl	47.15 g
$MgCl_2 \cdot 6H_2O$	18.10 g
$MgSO_4 \cdot 7H_2O$	7.00 g
$CaCl_2 \cdot 2H_2O$	3.13 g
Na_2SO_4	3.24 g
KCl	1.2 g
Na_2CO_3	0.1 g
NaBr	0.1 g
KBr	80.0 mg
$SrCl_2 \cdot 6H_2O$	72.0 mg
H_3BO_3	52.0 mg
Na_2HPO_4	8.1 mg
NaF	2.4 mg
Sodium silicate	0.4 mg
KI	50.0 µg
Distilled H_2O	1000.0 ml

DSM 623. ***Thermococcus litoralis*** **Medium**

Bacto-peptone (Difco 0123-17-3)	5.0 g
Yeast extract (Difco 0127-17-9)	1.0 g
NaCl	19.45 g
$MgCl_2 \cdot 6H_2O$	12.6 g
Na_2SO_4	3.42 g
$CaCl_2 \cdot 2H_2O$	2.38 g
KCl	0.55 g
Na_2CO_3	0.61 g
KBr	0.08 g
SrCl	0.057 g
Borsre	0.022 g
Na-meta-silicale	0.004 g
Na-fluoride	0.0024 g
KNO_3	0.0016 g
Na_2HPO_4	0.01 g
Sulfur	10.0 g
Resazurin	0.1 g
Distilled H_2O	1000.0 ml

Adjust final pH to 6.5. N_2 gassing, reduction with 0.5 g of Na_2S, and pH 6.5.

REFERENCES

Balch, W.E. and R.S. Wolfe. 1976. *Appl. Environ. Microbiol.* **32:** 781.

Balch, W.E., G.E. Fox, L.J. Magrum, C.R. Woese, and R.S. Wolfe. 1979. Methanogens: Reevaluation of a unique biological group. *Microbiol. Rev.* **43:** 260–296.

Bazylinski, D.A., C.O. Wirsen, and H.W. Jannasch. 1989. Microbial utilization of naturally-occurring hydrocarbons at the Guaymas Basin hydrothermal vent site. *Appl. Environ. Microbiol.* **55:** 2832–2836.

Brock, T.D., K.M. Brock, R.T. Belly, and R.L. Weiss. 1972. *Sulfolobus:* A new genus of sulfur-oxidizing bacteria living at low pH and high temperature. *Arch. Microbiol.* **84:** 54–68.

Darland, G., T. Brock, W. Samsonoff, and S. Conti. 1970. A thermophilic, acidophilic mycoplasma isolated from a coal refuse pile. *Science* **170:** 1416–1418.

Grogan, D. 1989. Phenotypic characterization of the archaebacterial genus *Sulfolobus:* Comparison of five wild-type strains. *J. Bacteriol.* **171:** 6710–6719.

Jones, W.J., J.A. Leigh, F. Mayer, C.R. Woese, and R.S. Wolfe. 1983. *Arch. Microbiol.* **136:** 254.

Kester, D.R., I.W. Duedall, D.N. Connors, and R.M. Pytkowicz. 1967. Preparation of artificial seawater. *Limnol. Oceanogr.* **12:** 176–178.

Patel, B.K.C., H.W. Morgan, and R.M. Daniel. 1985. *Fervidobacterium nodosum* gen. nov. and spec. nov., a new chemoorganotrophic, caldoactive, anaerobic bacterium. *Arch. Microbiol.* **141:** 63–69.

Pledger, R.J. and J.A. Baross. 1989. Characterization of an extremely thermophilic archaebacterium isolated from a black smoker polychaete (*Paralvinella* sp.) at the Juan de Fuca Ridge. *Syst. Appl. Microbiol.* **12:** 249–256.

Stetter, K.O. 1989. *Methanothermaceae.* In *Bergey's manual of systematic bacteriology* (ed. J.T. Staley et al.), vol. 3, pp. 2183–2184. Williams & Wilkins, Baltimore.

Wolin, E.A., M.J. Wolin, and R.S. Wolfe. 1963. Formation of methane by bacterial extracts. *J. Biol. Chem.* **283:** 2882–2886.

Zeikus, J.G. 1977. The biology of methanogenic bacteria. *Bacteriol. Rev.* **41:** 514–541.

Zeikus, J.G., P.W. Hegge, and M.A. Anderson. 1979. *Thermoanaerobium brockii* gen. nov. and spec. nov., a new chemoorganotrophic caldoactive, anaerobic bacterium. *Arch. Microbiol.* **122:** 41–48.

REFERENCES

Balch, W.E. and R.S. Wolfe. 1976. *Appl. Environ. Microbiol.* **32:** 781.

Balch, W.E., G.E. Fox, L.J. Magrum, C.R. Woese, and R.S. Wolfe. 1979. Methanogens: Reevaluation of a unique biological group. *Microbiol. Rev.* **43:** 260–296.

Bazylinski, D.A., C.O. Wirsen, and H.W. Jannasch. 1989. Microbial utilization of naturally-occurring hydrocarbons at the Guaymas Basin hydrothermal vent site. *Appl. Environ. Microbiol.* **55:** 2832–2836.

Brock, T.D., K.M. Brock, R.T. Belly, and R.L. Weiss. 1972. *Sulfolobus:* A new genus of sulfur-oxidizing bacteria living at low pH and high temperature. *Arch. Microbiol.* **84:** 54–68.

Darland, G., T. Brock, W. Samsonoff, and S. Conti. 1970. A thermophilic, acidophilic mycoplasma isolated from a coal refuse pile. *Science* **170:** 1416–1418.

Grogan, D. 1989. Phenotypic characterization of the archaebacterial genus *Sulfolobus:* Comparison of five wild-type strains. *J. Bacteriol.* **171:** 6710–6719.

Jones, W.J., J.A. Leigh, F. Mayer, C.R. Woese, and R.S. Wolfe. 1983. *Arch. Microbiol.* **136:** 254.

Kester, D.R., I.W. Duedall, D.N. Connors, and R.M. Pytkowicz. 1967. Preparation of artificial seawater. *Limnol. Oceanogr.* **12:** 176–178.

Patel, B.K.C., H.W. Morgan, and R.M. Daniel. 1985. *Fervidobacterium nodosum* gen. nov. and spec. nov., a new chemoorganotrophic, caldoactive, anaerobic bacterium. *Arch. Microbiol.* **141:** 63–69.

Pledger, R.J. and J.A. Baross. 1989. Characterization of an extremely thermophilic archaebacterium isolated from a black smoker polychaete (*Paralvinella* sp.) at the Juan de Fuca Ridge. *Syst. Appl. Microbiol.* **12:** 249–256.

Stetter, K.O. 1989. *Methanothermaceae.* In *Bergey's manual of systematic bacteriology* (ed. J.T. Staley et al.), vol. 3, pp. 2183–2184. Williams & Wilkins, Baltimore.

Wolin, E.A., M.J. Wolin, and R.S. Wolfe. 1963. Formation of methane by bacterial extracts. *J. Biol. Chem.* **283:** 2882–2886.

Zeikus, J.G. 1977. The biology of methanogenic bacteria. *Bacteriol. Rev.* **41:** 514–541.

Zeikus, J.G., P.W. Hegge, and M.A. Anderson. 1979. *Thermoanaerobium brockii* gen. nov. and spec. nov., a new chemoorganotrophic caldoactive, anaerobic bacterium. *Arch. Microbiol.* **122:** 41–48.

Appendix 3

Chromosome Map of *Thermococcus celer* Strain Vu13

Contributed by K.M. Noll

The *Thermococcus celer* chromosome map was generated using partial and complete digests by restriction endonucleases *Nhe*I, *Spe*I, and *Xba*I. Digests were resolved by pulsed-field gel electrophoresis. Fragments were aligned through analyses of partial digests and hybridization of labeled restriction fragments to resolved fragments. The genome consists of a single, circular chromosome. The genes for ribosomal RNAs were located on the map by hybridization. The 16S–23S cluster lies at the junction of fragments S2 and S4a with the *Spe* site near the 5′ end of the 23S gene. The 5S rRNA gene lies between the N3-N5 and X4-X3 junctions, and the 7S RNA gene lies between the X1-X2 and S4b-S1 junctions. The 0-kb map position is drawn so that the direction of transcription of the 23S rRNA gene is clockwise. Fragments X9, X10, and X11 are tentatively assigned the positions shown here. Details of the construction of the map can be found in Noll (*J. Bacteriol. 171:* 6720–6725 [1989]).

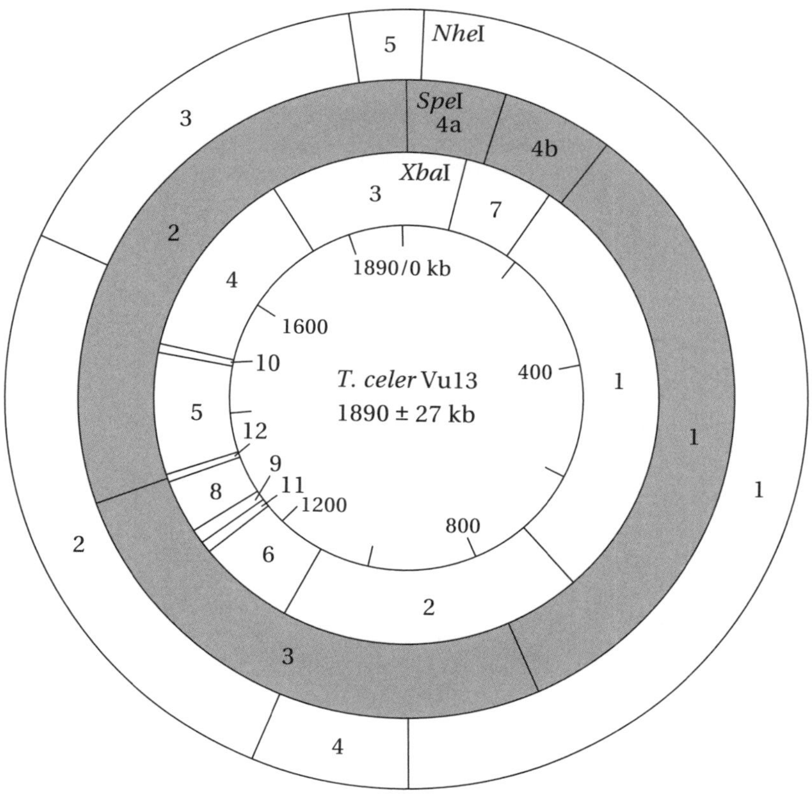

Appendix 4

Codon Usage Tables for Thermophilic Archaea

Compiled by J. DiRuggiero, K.M. Borges, and F.T. Robb

The table clearly shows that the universality of the genetic code is conserved in the Archaea. However, G and C are rarely used in the third positions of codons, and a strong bias against the CG dinucleotide is observed. This bias is reflected in the arginine codon usage, with six possibilities; AGG and AGA are strongly preferred and the other four codons are rarely found. The opposite bias occurs in *Escherichia coli*, where the codons CGA, CGT, CGG, and CGC are used in preference to AGG and AGA. The bias against CG is also found for alanine, proline, and serine codons in thermophiles, but to a lesser extent than for arginine codons. The CG dinucleotide-containing codons are frequently used for alanine, proline, and serine in *E. coli* (Zhang et al. 1991).

Cysteine codons are underepresented in thermophiles, which may reflect the relative instability of this amino acid at high temperature. The stop codons in thermophilic Archaea show a possible bias in favor of TGA, although more data are required to confirm this observation. The codon usage of vertebrates, such as rabbits, does not show a bias against codons preferentially used by hyperthermophilic Archaea.

Codon usage is useful in designing oligonucleotides from amino acid sequences, in detecting open reading frames in genomic sequence data, and in selecting host systems for heterologous expression of archaeal genes.

The paper by Wada et al. (1992) contains data representing all of the organisms for which more than 20 genes are available and thus can be used for comparative studies of archaeal codon usage.

REFERENCES

Wada, K., Y. Wada, F. Ishibashi, T. Gojobori, and T. Ikemura. 1992. Codon usage tabulated from GenBank genetic sequence data. *Nucleic Acids Res.* **20:** 2111–2118.

Zhang, S., G. Zubay, and E. Goldman. 1991. Low-usage codons in *Escherichia coli*, fruit fly and primates. *Gene* **105:** 61–72.

Codon Usage Table for Archaea

Amino acid	Codon	ES4	*Pyrococcus furiosus*	*Thermococcus litoralis*	*Thermococcus celer*	*Sulfolobus acidocaldarius*	*Halobacterium* sp.	*Methanococcus vaniellii*	*Escherichia coli*	Rabbit
Ala	GCG	7.1	4.3	3.3	15.4	4.5	40.2	2.9	33.1	8.4
	GCA	35.6	30.8	23.9	4.3	26.0	11.8	48.9	20.6	11.5
	GCT	28.5	27.4	17.8	3.7	22.0	8.1	30.0	17.4	15.3
	GCC	21.4	13.1	13.1	44.6	6.6	41.7	2.6	23.5	33.9
Arg	AGG	14.3	20.3	21.1	44.6	15.4	0.6	10.6	1.4	10.7
	AGA	28.5	36.8	24.4	8.6	44.6	1.1	27.1	2.1	8.3
	CGG	0.0	0.0	2.3	3.1	0.2	14.8	1.5	4.6	12.1
	CGA	2.4	0.0	1.9	0.3	0.5	8.0	1.4	3.1	5.9
	CGT	2.4	1.4	1.9	4.9	0.9	6.3	0.9	24.1	3.9
	CGC	0.0	0.6	4.2	16.3	0.0	28.4	0.3	22.1	14.1
Asn	AAT	4.8	10.3	15.5	0.9	30.7	2.6	15.2	16.3	12.5
	AAC	30.9	22.3	25.8	32.3	12.5	27.9	25.4	23.9	25.4
Asp	GAT	26.1	27.4	26.7	11.1	43.6	14.2	28.3	32.3	18.5
	GAC	30.9	24.9	26.7	42.4	12.0	78.4	26.8	21.8	34.7
Cys	TGT	2.4	2.3	3.7	0.3	3.3	2.0	7.0	4.7	6.4
	TGC	0.0	0.9	2.3	5.5	0.7	2.1	2.6	6.1	11.6
Gln	CAG	21.4	14.6	6.1	26.7	11.6	27.0	11.2	30.1	32.2
	CAA	7.1	12.6	9.8	0.3	17.5	6.3	15.7	13.2	8.3
Glu	GAG	66.5	42.6	43.1	79.9	26.7	64.1	6.5	19.2	48.0
	GAA	16.6	49.4	49.2	9.8	46.8	29.4	79.6	43.4	22.7
Gly	GGA	42.8	34.8	24.8	17.2	30.2	6.3	31.9	7.0	14.8
	GGT	35.6	22.9	18.7	27.4	33.5	12.3	30.5	27.6	8.3
	GGC	4.8	6.3	16.4	22.4	6.3	41.7	7.7	30.3	27.8
	GGG	0.0	5.4	8.0	9.8	4.3	16.2	7.5	9.7	17.8
His	CAT	0.0	4.9	8.0	1.2	10.8	2.1	5.3	11.6	7.1
	CAC	11.9	15.4	4.7	22.4	4.7	18.0	11.6	10.7	17.5
Ile	ATA	30.9	30.0	36.1	30.1	51.7	0.9	15.2	4.1	5.2
	ATT	23.8	36.8	23.0	4.3	30.5	3.5	44.6	27.2	13.6
	ATC	19.0	19.1	19.7	39.3	5.7	39.0	21.6	26.5	30.9
Leu	TTG	0.0	6.6	8.9	1.2	12.0	6.2	7.5	11.5	9.2
	TTA	2.4	10.6	12.7	0.9	43.6	1.7	37.8	10.9	4.3
	CTG	0.0	2.3	8.4	12.6	2.6	24.6	0.5	54.6	50.9
	CTA	7.1	12.9	6.6	2.5	12.3	2.7	3.7	3.2	4.5
	CTT	23.8	21.1	28.6	11.1	13.2	5.1	17.5	10.2	9.1
	CTC	26.1	20.0	18.7	58.1	2.8	40.9	5.8	9.9	24.4

Lys	AAG	78.4	56.3	50.6	57.8	26.0	18.4	14.8	12.0	36.4
	AAA	2.4	26.6	52.0	8.3	42.5	9.5	72.2	36.5	18.4
Met	ATG	30.9	27.1	20.6	29.5	25.9	19.8	27.6	26.5	24.9
Phe	TTT	7.1	14.0	26.7	2.2	18.0	2.1	19.3	19.2	16.4
	TTC	16.6	18.6	15.5	25.8	11.1	25.5	12.6	18.2	32.1
Pro	CCG	0.0	1.1	5.6	27.7	2.8	22.2	3.6	23.8	8.9
	CCA	38.0	34.8	17.3	4.6	21.0	4.3	20.3	8.2	12.0
	CCT	7.1	6.3	5.6	2.5	18.9	2.7	14.8	6.6	13.2
	CCC	0.0	7.1	8.9	17.5	6.4	13.6	2.4	4.3	22.5
Ser	AGT	2.4	9.4	7.5	0.3	16.5	4.2	8.5	7.2	7.3
	AGC	21.4	12.6	14.5	20.0	5.9	14.1	5.3	15.2	19.4
	TCG	0.0	0.9	1.9	5.5	1.6	18.4	2.0	8.0	4.4
	TCA	2.4	6.3	10.3	1.2	16.0	3.1	23.0	6.8	6.7
	TCT	0.0	3.4	8.0	1.5	14.6	1.9	6.5	10.4	9.0
	TCC	4.8	3.7	4.2	10.5	4.5	16.1	4.1	9.4	18.7
Thr	ACG	4.8	1.7	7.0	12.3	3.3	29.6	4.8	12.7	8.6
	ACA	16.6	18.9	15.0	3.1	17.5	6.3	17.2	6.5	10.5
	ACT	7.1	17.4	12.2	1.2	22.2	4.2	16.7	10.2	9.4
	ACC	26.1	6.3	9.8	33.8	5.6	24.4	7.5	24.3	22.7
Trp	TGG	23.8	12.3	15.5	7.1	7.6	8.8	4.4	12.8	13.5
Tyr	TAT	9.5	15.7	25.8	2.8	28.6	2.2	11.8	15.4	9.5
	TAC	40.4	24.6	24.8	28.0	9.5	22.7	17.9	13.4	20.7
Val	GTG	4.8	9.4	10.3	12.0	9.4	27.0	3.1	25.3	32.9
	GTA	9.5	24.0	17.3	2.2	33.1	3.5	31.9	11.6	4.5
	GTT	47.5	36.3	33.3	24.0	31.8	7.7	48.4	20.1	8.6
	GTC	21.4	11.7	12.7	40.9	6.9	47.5	2.7	14.2	18.4
End	TGA	2.4	1.1	0.9	1.8	1.0	2.1	0.3	0.8	0.9
	TAG	0.0	0.3	0.0	0.3	0.0	0.5	0.3	0.2	0.4
	TAA	0.0	0.6	0.0	0.0	1.4	1.6	3.1	2.0	0.5
Total number of codons		421	3,501	2,134	3,253	5,764	17,967	5,869	524,410	100,704

Frequency of use per 1000 codons.

Codon Usage Table for Archaea

DNA sequence	GenBank number	Reference
ES4		
glutamate dehydrogenase	L12408	DiRuggiero et al. (1993)
P. furiosus		
aldehyde oxidoreductase	–	M.W.W. Adams (unpubl.)
DNA polymerase	D12983 (DDBJ database)	Uemori et al. (1993)
elongation factor 1α	X59857	Creti et al. (1991)
elongation factor 2	X67205	R. Creti (unpubl.)
transcription initiation factor IIB	X70668	R. Creti (unpubl.)
ferredoxin	–	M.W.W. Adams (unpubl.)
glutamate dehydrogenase	–	Eggen et al. (1993)
glutamine synthetase	L12410	Brown et al. (1994)
glyceraldehyde 3-phosphate dehydrogenase	M83988	Zwickl et al. (1990)
S10 ribosomal protein	X59857	Creti et al. (1991)
T. litoralis		
DNA polymerase	M74198	Perler et al. (1992)
endonuclease *Tli*I	M74198	Perler et al. (1992)
endonuclease *Tli*II	M74198	Perler et al. (1992)
glutamate dehydrogenase	L19115	K.M. Borges et al. (pers. comm.)
T. celer		
elongation factor 1α	X52383	Auer et al. (1990)
ribosomal protein 30	X67313	Klenk et al. (1992b)
ribosomal protein S7	X67313	Klenk et al. (1992b)
RNA polymerase (subunits A1, A2, B)	X67313	Klenk et al. (1992b)
RNA polymerase (subunit H)	X59077	Klenk et al. (1992a)
Sulfolobus acidocaldarius	–	Wada et al. (1992)
Halobacterium spp.	–	Wada et al. (1992)
Methanococcus vannielii	–	Wada et al. (1992)
Escherichia coli	–	Wada et al. (1992)
Rabbit	–	Wada et al. (1992)

References:

Auer, U., G. Spicker, and A. Böck. 1990. Nucleotide sequence of the gene for elongation factor EF-1α from the extreme thermophilic archaebacterium *Thermococcus celer. Nucleic Acids Res.* **89:** 3989.

Brown, J.R., Y. Masuchi, F.T. Robb, and W.F. Doolittle. 1994. Evolutionary relationships of bacterial and archaea glutamine synthetase. *J. Mol. Evol.* **38:** 566–576.

Creti, R., F. Citarella, O. Tiboni, A.M. Sanangelantoni, P. Palm, and P. Cammaramo. 1991. Nucleotide sequence of a DNA region comprising the gene for elongation factor 1α (EF-1α) from the ultrathermophilic archaeote *Pyrococcus woesei*: Phylogenic implications. *J. Mol. Evol.* **33:** 332–342.

DiRuggiero, J., F.T. Robb, R. Jagus, H.H. Klump, K.M. Borges, M. Kessel, X. Mai, and M.W.W. Adams. 1993. Characterization, cloning, and in vitro expression of the extremely thermostable glutamate dehydrogenase from the hyperthermophilic archaeon, ES4. *J. Biol. Chem.* **268:** 17767–17771.

Eggen, R.I.L., A.C.M. Geerling, K. Waldkotter, G. Antranikian, and W.M. de Vos. 1993. The glutamate dehydrogenase-encoding gene of the hyperthermophilic archaeon *Pyrococcus furiosus:* Sequence, transcription, and analysis of the deduced amino acid sequence. *Gene* **132:** 143–148.

Klenk, H.P., P. Palm, F. Lottspeich, and W. Zillig. 1992a. Component H of the DNA-dependent RNA polymerases of archaea is homologous to a subunit shared by the three eucaryal nuclear RNA polymerases. *Proc. Natl. Acad. Sci.* **89:** 407–410.

Klenk, H.P., V. Schwass, F. Lottspeich, and W. Zillig. 1992b. Nucleotide sequence of the genes encoding the three largest subunits of the DNA-dependent RNA polymerase from the archaebacterium *Thermococcus celer. Nucleic Acids Res.* **20:** 4659–4659.

Perler, F.B., D.G. Comb, W.E. Jack, L.S. Moran, B. Qiang, R.B. Kucera, J. Benner, B.E. Slatko, D.O. Nwankwo, S.K. Hempstead, C.K.S. Carlow, and H. Jannasch. 1992. Intervening sequences in an archaea DNA polymerase gene. *Proc. Natl. Acad. Sci.* **89:** 5577–5581.

Uemori, T., Y. Ishino, H. Toh, K. Asada, and I. Kato. 1993. Organization and nucleotide sequence of the DNA polymerase gene from the archaeon *Pyrococcus furiosus. Nucleic Acids Res.* **21:** 259–265.

Wada, K., Y. Wada, F. Ishibashi, T. Gojobori, and T. Ikemura. 1992. Codon usage tabulated from GenBank genetic sequence data. *Nucleic Acids Res.* **20:** 2111–2118.

Zwickl, P., S. Fabrey, C. Bogedain, A. Haas, and R. Hensel. 1990. Glyceraldehyde-3-phosphate dehydrogenase from the hyperthermophilic archaebacterium *Pyrococcus woesei*: Characterization of the enzyme, cloning and sequencing of the gene, and expression in *Escherichia coli. J. Bacteriol.* **172:** 4329–4338.

Appendix 5

The Berlin RNA Databank: Compilation of 5S rRNA and 5S rRNA Gene Sequences

Contributed by T. Specht and V.A. Erdmann

The Berlin RNA Databank was initiated in 1977 when 13 prokaryotic and 23 eukaryotic 5S rRNA sequences were known (Erdmann 1977). The Berlin RNA Databank as of January, 1994, contains a total of 733 sequences of 5S rRNAs or their genes. It includes sequences from 44 Archaea, 309 eubacteria, 20 plastids, 6 mitochondria, 333 eukaryotes, and 21 eukaryotic pseudogenes. The databank format used is the format of the EMBL nucleotide sequence data library complemented by a Sequence Alignment (SA) field including secondary structural information. The databank is also available from the EMBL in Heidelberg, Germany, on CD-ROM together with other databases (EMBL Nucleotide) or online via E-mail ftp (ftp.embl-heidelberg.de/pub/databases/berlin). Furthermore, the data are available from us by request on 3 1/2" or 5 1/4" diskettes for IBM or compatibles (requests by mail or E-mail to thy@chemie.fu-berlin.de). In addition to the standard database entries, we are also including the sequences in an aligned format introducing a standard numeration based on the sequence of *Escherichia coli* 5S rRNA. The secondary structural information is displayed according to the minimal 5S rRNA structural model (Specht et al. 1990, 1991a,b). The table shows the overall alignment of all archaeal 5S rRNAs including secondary structure information. Alignment gaps are indicated by hyphen (–). Sequence heterogeneity deduced from the autoradiogram without sequencing distinct bands separately is indicated as recommended by the IUB-IUPAC joint commission in biochemical nomenclature (Cornish-Bowden 1985): R=AG, Y=CU, M=AC, S=GC, W=AI, B=notA, D=notC, H=notG, V=notU. Incomplete sequence analysis is indicated by X. DNA sequences are indicated by T instead of U, pseudouridine by F, length heterogeneity by lowercase letters. The numeration is according to the *E. coli* sequence (Erdmann 1977). ([]) Helix boundaries of the secondary structure and bulge boundaries within helices; (!) helix end and start of a new helix at the same position; (<>) odd base-pairs are parenthesized by less/greater symbols.

The secondary structures of all types of 5S rRNAs consist of five helices, named A to E, which are connected by loops designated as *a* to *e* as shown in Figure 1. The Helix D is best suited to distinguish between the five different 5S rRNA structural groups (Fig. 1). One group consists of eukaryotes and one of eubacteria, including plastids and mitochondria. For Archaea, we distinguish a total of three different groups, division I (comprising Thermococcales, Thermoplasmatales, Methanobacteria and Halobacteria), division II (Eocyta), and the last group is represented by the single organism, Octopus Spring species I (Specht et al. 1990, 1991a,b).

Alignment of All 44 Archaeal 5S rRNA Sequences

```
              10         20          30         40         50          60         70         80          90        100
               .          .           .          .          .           .          .          .           .          .
1 - - - - -[G C C C - A - - C C C G G U]C - A C - A[G U - G A - G C G]G G C - A A C[A C - C c G G]A C  SULFOLOBUS ACIDOCALDARIUS +
1 - - - - -[G C C C - A - - C C C G G C]C - A C - A[G T - G A - G C G]G G C - A A C[A C - C C G G]A C  SULFOLOBUS B12
1 - - - - - - -[C C - A - - C C C G G C]G - A U - A[G U - G<A>- G<C>G]G G C - A A C[A C - C C G G]A C  SULFUROCOCCUS MIRABILIS
1 A C[G G T G C C C G A - - C C C G G C]C - A T - A[G T - G G - C C G]G G C - A A C[A C - C C G G]T C  DESULFUROCOCCUS MOBILIS
1 - - -[U G G C C C G A - - C C C G G C]C - A U - A[G C - G G - C C G]G G C - A A C[A C - C C G G]A C  PYRODICTIUM OCCULTUM
1 - - - - G[C C C C - U[A A]C C C G G C]A - A U - A[G G C G C - C G G]U G C - U A C[G C - C C G G]U C  OCTOPUS SPRING SPECIES 1
1 - - - - -[G G - U[A]C - - G G C G G U]C - A U - A[G C - G G - C G G]G G U - C A C[A C - C C G G]U C  THERMOCOCCUS CELER [XTR]
1 - - - - -[G G - U[A]C - - G G C G G U]C - A U - A[G C - G G - G G G]G G C - C A C[A C - C C G G]U C  PYROCOCCUS WOESEI
1 - - - - -[G G - - -<C>- - A A C G G U]C - A U - A[G C - A G - C A G]G G A - A A C[A C - C<A>G A]U C  THERMOPLASMA ACIDOPHILUM
1 - - - -[G G G - T[A]C - - G G C G A C]C - A T - A[G C - G G - G G G]G G C - C A C[A C - C T G A]A C  METHANOTHERMUS FERVIDUS
1 - - - -[A S G - U - U - - G G C G G C]C - A U - A[G C - G G - A G G]G G C - C A U[A C - C U G G]U C  METHANOBACT. FORMICICUM
1 - - - U[A G G - U[U]U - - G G C G G U]C - A U - A[G C - G A -<U>G G]G G U - U A C[A C - C U G G]U C  METHANOBREVIBACTER RUMINANTIUM
1 - - - G G[G G - C[A]U - - G G C G G U]C - A U - G[G C - G C - A G G]G G U - U A U[A C - C U G A]U C  MBACT. THERMOAUTOTROPHICUM A1
1 - - - - G[G G U U[U]U - - G G C G G U]C - A U - G[G C - G U - G G G]G G U U U A U[A C - C U G A]U C  MBACT. THERMOAUTOTROPHICUM A2
1 - - - - - -[G - U - U - - G G C G G U]C - A U - G[G C - G U - G G G]G U U - U A U[A C - C U G A]U C  MBACT. THERMOAUTOTROPHICUM B
1 - - - - - -[G - U - U - - G G C G G U]C - A U - G[G C - G U - G G G]G U U - A A U[A C - C U G A]U C  METHANOBACT. THERMOFORMICICUM
1 - - - -[U G A - U[A]U - - A A C G G U]C - A U - A[G C - G G - A G G]A G C - C A C[A U - C C G A]U C  MCOCCUS THERMOLITHOTROPHICUS
1 - - - -[T G A - C[A]T - - A A C G G T]C - A A - A[G C - G G - A G G]T G T - A A C[A T - C C G A]T C  METHANOCOCCUS VANNIELII [LNK]
1 - - - -[T G A - C[A]T - - A A C G G T]C - A T - A[G C - G G - A G G]T G T - A A C[A T - C C G A]T C  METHANOCOCCUS VOLTAE [LNK]
1 - - - -[T G A - T[A]C - - G G C G G T]C - A T - A[G C - G G - G G G]T G T - A A C[A T - C C G A]T C  METHANOCOCCUS VANNIELII [XTR]
1 - - - -[T G A - T[A]C - - G G C G G T]C - A T - A[G C - G G - A G G]T G T - C C C[A T - C C G A]T C  METHANOCOCCUS VOLTAE [XTR]
1 - - - - U[C A - A - U - - A G C G G C]C - A C - A[G<C>- A G - G U G]U G U - C A C[A C - C C G U]U C  METHANOSPIRILLUM HUNGATII
1 - - - -[G G A G<U[U]U - - G G C G G C]C - A U - A[G C - G G - C G G]G G C - A A U[U C - C U G U]A C  METHANOLOBUS TINDARIUS A
1 - - - - X[G - G - - U - - G G C G G C]C - A U - A[G C - G G - C G G]G G C - A A U[U C - C U G U]A C  METHANOLOBUS TINDARIUS B
1 - - - - -[G A - U<C>U - - G G C A[G]C]C - A U - A[G C - G G - C A G]G U G - U A A[U C - C U G U]A C  METHANOSARCINA VACUOLATA
1 - - - - G[G A G U<U>U - - G G C G - C]C - A U - A[G C - G G - C A G]U G U - A A C[U C - C U G U]A C  METHANOSARCINA BARKERI
1 - - - - - - X X[U - U - - G G C G G U]C - A U - G[A C - G G - C A G]G G U - A A C[U C - C U G U]A C  METHANOSARCINA ACETIVORANS
1 - - - - - - A[G C - U - - A G U G G C]C - A U - A[G U - G G - C A G]G G A - C C C[U C - C U G U]U C  METHANOCOCCOIDES METHYLUTENS
1 - - - - - - X X[U - U - - G G C G G C]C - A U - G[G C - G G - C G G]G G U - A A C[U C - C U G U]A C  'METHANOCOCCUS' HALOPHILUS
1 - - - - - - X X[U - U - - A G C G G C]C - A U - G[G C - G G - C A G]G U U - A A C[U C - C U G U]A C  METHANOHALOBIUM EVESTIGATUS
1 - - - - - - -[G A - A - - G G C G G C]C - A G - A[G C - G G - U A G]G G A - A A C[A C - C C G U]A C  HALOBACT.DISTRIBUTUS BKM 4
1 - - - - - - -[G U - A - - G G C G G C]C - A A - A[G C - G G - C A G]G G A - A A G[A C - C C G U]A C  HALOBACT.DISTRIBUTUS BKM13 [U]
1 - - - - - - - X U -[A - - G G C G G C]C - G U - A[G C - G G - C G G]G U G - U C C[U C - C C G U]A C  HALOBACT.DISTRIBUTUS BKM13 [M]
1 - - - - - - U U A - A - -[G G C G G C]C - A G - A[G C - G G - U G A]G G U - U C C[A C - C C G U]A C  HALOFERAX VOLCANII
1 - - - - - A A U A -[G - - G G C G G C]C - A G - A[G C - G G - U G A]G G U - U C C[A C - C C G U]A C  HALOFERAX MEDITERRANEI
1 - - - - - - - X[U - - - - G G C G G C]U - A U - A[G C - G G - C G G]G G A - A A C[A C - C C G U]A C  HALOCOCCUS TURKMENICUS
1 - - - - - - U U A -[A - - G G C G G C]C - A C - A[G C - G G - C G G]G G C - G A C[U C - C C G U]A C  HALOCOCCUS MORRHUAE
1 - - - - - - T T A - - - -[G G C G G C]C - A C - A[G C - G G - T G G]G G T T G C C[T C - C C G T]A C  HALOARCULA MARIS-MORTUI
1 - - - - - - U U A -[A - - G G C G G C]C - A U - A[G C - G G - U G G]G G U - U A C[U C - C C G U]A C  HALOBACTERIUM SALINARIUM
1 - - - - - - -[U A - A - - G G C G G C]C - A U - A[G C - G G - C G G]G G U - U C C[U C - C C G U]A C  NATRONOCOCCUS OCCULTUS
1 - - - - - - - -[A - A - - G G C G G C]C - A U - A[G C - G G - C G G]G G U - C C C[U C - C C G U]A C  NATRONOBACT. PHARAONIS BKM 6
1 - - - - - - - -[A - A - - G G C G G C]C - A U - A[G C - G G - C G G]G G U - C C C[U C - C C G U]A C  NATRONOBACT. PHARAONIS BKM 12
```

```
               110       120       130       140       150       160       170       180       190       200
                .         .         .         .         .         .         .         .         .         .
37 - - U C A U U U C G A A C[C C G G[A - A - - - -]G U]U A A G C[C G C U C - - A C]G -[U U A G U G G -  SULFOLOBUS ACIDOCALDARIUS +
37 - - T C A T T T C G A A C[C C G G[A - A - - - -]G T]T A A G C[C G C T C - - A C]G -[T T G G T G G -  SULFOLOBUS B12
35 - - U C A U C U C G A A C[C C G G[A - A - - - -]G U]A A A G U[C<C>C<C>U - - A C]G -[U U G G U A A -  SULFUROCOCCUS MIRABILIS
43 - - T C G T T T C G A A C[C C G G[A - A - - - -]G T]T A A G C[C G G C C - - A C]G -[T C A G A A C -  DESULFUROCOCCUS MOBILIS
40 - - U C A U G U C G A A C[C C G G[A - A - - - -]G U]U A A G C[C G G C C - - G C]G -[U U G G G G G[A  PYRODICTIUM OCCULTUM
41 - - U C U U - C A G A A C[C C G G[A - A - - - -]G C]U A A G G[C C G G C[- G]C C]G -[C G G[A]C G G]G  OCTOPUS SPRING SPECIES 1
37 - - U C G U U U C G A C C[C C G G[A - A - - - -]G U]U A A G C[C C G C C[- A]G C]G - A[U C C C G G]U  THERMOCOCCUS CELER [XTR]
37 - - U C A U U U C G A A C[C C G G[A - A - - - -]G U]U A A G C[C C C C C[- A]G C]G - A[U C C C G G]C  PYROCOCCUS WOESEI
35 - - C C A U U C C G A A C[U C[G A]C>- G - - - - G U]U A A G C[C U G C U - - G C]G -[U A U U G C G]U  THERMOPLASMA ACIDOPHILUM
38 - - C C G T T C C G A T C[T C A G[A - A - - - -]G T]T A A G C[C C C C C[- T]G C]G -[T T G G A G G]T  METHANOTHERMUS FERVIDUS
37 - - U C G U U U C G A U C[C C A G[A - A - - - -]G U]G A A G U[C C U C C[- U]G C]G - U U U[U G]U U G  METHANOBACT. FORMICICUM
39 - - U C G U U U C G A U C[C C A G[A - A - - - -]G U]U A A G U[C U<U>U U[- C]G C]G - U U U[U G]U U U  METHANOBREVIBACTER RUMINANTIUM
39 - - U C G U U U C G A U C[U C A G[U - A - - - -]G U]U A A G U[C C U G C[- U]G C]G - U U[G U G G G]U  MBACT. THERMOAUTOTROPHICUM A1
40 - - U C G U U U C G A U C[U C A G[U - A - - - -]G U]U A A G U[C C U G C[- U]G C]G - U U[G U G G G]U  MBACT. THERMOAUTOTROPHICUM A2
35 - - U C G U U U C G A U C[U C A G[U - A - - - -]G U]U A A G U[C C U G C[- U]G C]G - U U[G U G G G]U  MBACT. THERMOAUTOTROPHICUM B
35 - - U C G U U U C G A U C[U C A G[U - A - - - -]G U]U A A G U[C C U G C[- U]G C]G - U U[G U G G G]U  METHANOBACT. THERMOFORMICICUM
38 - - C C G U C C C G A U C[U C G G[A - A - - - -]A U]U A A G C[C C U C C[- A]G C]G - A[U A C C C U]A  MCOCCUS THERMOLITHOTROPHICUS
38 - - C C A T C C C G A T C[T C G G[A - A - - - -]A T]T A A G C[C C T C C[- A]G C]G - A T T[C T T T]A  METHANOCOCCUS VANNIELII [LNK]
38 - - C C A T T C C G A T C[T C G G[A - A - - - -]A T]T A A G C[C C T C C[- A]G C]G - A T T[T C T T]A  METHANOCOCCUS VOLTAE [LNK]
38 - - C C A T C C C G A T C[T C G G[A - A - - - -]A T]T A A G C[C C T C C[- A]G C]G - A T T[C C T T]A  METHANOCOCCUS VANNIELII [XTR]
38 - - C C A T T C C G A T C[T C G G[A - A - - - -]A T]T A A G C[C C T C C[- A]G C]G - A T T[T C T T]A  METHANOCOCCUS VOLTAE [XTR]
37 - - C C A U U C C G A A C[A C G G[A - A - - - -]G U]U A A G A[C A C C U[- C]A>C]G - U[G G A U G A]C  METHANOSPIRILLUM HUNGATII
39 - - C C A U C C C G A A C[A C A G[A - A - - - -]G A]U A A G C[C C G C C[- U]G C]G -[U U C U A U A]C  METHANOLOBUS TINDARIUS A
36 - - C C A U C C C G A A C[A C A G[A - A - - - -]G A]U A A G C[C C G C C[- U]G C]G -[U U C U A U A]C  METHANOLOBUS TINDARIUS B
37 - - C C A U C C C G A A C[A C A G[U - A - - - -]G A]U A A G C[C U G C C[- C]G C]G - U[U C C]U U G C  METHANOSARCINA VACUOLATA
38 - - C C A U C C C G A A C[A C A G[U - A - - - -]G A]U A A G C[C U G C C[- C]G C]G - U[U C C]U U A C  METHANOSARCINA BARKERI
36 - - C C A U U C C G A A C[A C A G[C - A - - - -]G A]U A A G C[C U G C C[- C]G U]G - U[U C C]U U A C  METHANOSARCINA BARKERI
36 - - A C C U U C C G A A C[A C A G[A - A - - - -]G A]U A A G C[C U G C C[- C]A C]G - U[U G C]U U A C  METHANOSARCINA ACETIVORANS
36 - - C C A U C C C G A A C[A C A G[C - A - - - -]G A]U A A A C[C C G C C[- C]G C]G - U[U C C G U A]C  METHANOCOCCOIDES METHYLUTENS
36 - - C C A U C C C G A A C[A C A G[A - A - - - -]G A]U A A G C[C U G C C[- A]G C]G - U U[G U A U A]C  'METHANOCOCCUS' HALOPHILUS
35 - - C C A U U C C G A A C[A C G G[A - A - - - -]G U]U A A G C[C U A C C[- A]G C]G - U A[U C G U G]A  METHANOHALOBIUM EVESTIGATUS
35 - - C C A U U C C G A A C[A C G G[A - A - - - -]G U]G A A G C[C U G C C[- A]G C]G - U[A C C A G C]U  HALOBACT.DISTRIBUTUS BKM 4
35 - - C C A U C C C G A A C[A C G G[A - A - - - -]G A]U A A G C[U C G C C[- U]G C]G - U C[C>C G G C]A  HALOBACT.DISTRIBUTUS BKM13 [U]
36 - - C C A U C C C G A A C[A C G G[A - A - - - -]G U]U A A G C[U C A C C[- U]G C]G - U U[C U G G]U C  HALOBACT.DISTRIBUTUS BKM13 [M]
37 - - C C A U C C C G A A C[A C G G[A - A - - - -]G U]U A A G C[U C G C C[- U]G C]G - U U[C U G G<U]C  HALOFERAX VOLCANII
34 - - U C A U C C C G A A C[A C G G[A - C - - - -]G U]U A A G C[C C G C C[- U]G C]G - U U[U C G G C]G  HALOFERAX MEDITERRANEI
36 - - C C A U C C C G A A C[A C G G[C - A - - - -]G A]U A A G C[C C G C C[- A]G C]G - U U[C C A G<C]G  HALOCOCCUS TURKMENICUS
36 - - C C A T C C C G A A C[A C G G[A - A - - - -]G A]T A A G C[C C A C C[- A]G C]G - T T[C C<A>G G]G  HALOCOCCUS MORRHUAE
36 - - C C A U C C C G A A C[A C G G[A - A - - - -]G A]U A A G C[C C G C C[- U]G C]G - U U[C C G G]U C  HALOARCULA MARIS-MORTUI
35 - - C C A U C C C G A A C[A C G G[A - A - - - -]G A]U A A G C[C C G C C[- U]G C]G - U[A U U G G U]G  HALOBACTERIUM SALINARIUM
34 - - C C A U C C C G A A C[A C G G[A - A - - - -]G A]U A A G C[C C G C C[- U]G C]G - U[A U U G G U]G  NATRONOCOCCUS OCCULTUS
34 - - C C A U C C C G A A C[A C G G[A - A - - - -]G A]U A A G C[C C G C C[- U]G C]G - U[A U U G G U]G  NATRONOBACT. PHARAONIS BKM 6
```

```
         210          220          230          240          250          260          270          280          290          300
          .            .            .            .            .            .            .            .            .            .
76  - G G C]- C[G U G G A U[A]C C]- G U G A -[G G A U C C G C]A -[G C C C C A - - - - C - - - U - A A!G   SULFOLOBUS ACIDOCALDARIUS +
76  - G G C]- C[G T G G A T[A]C C]- G T G A -[G G A T C C G C]A -[G C C C C A - - - - C - - - T - A A!G   SULFOLOBUS B12
74  - G G C]- A[G U G G G<A[U]C C]- G C A A -[G G<G>C C U G C]A -[G C C U U G - - - - C - - - C - A A!G   SULFUROCOCCUS MIRABILIS
82  - G G C]- C[G T G A G G[T]C C]- G A G A -[G G C C T C G C]A -[G C C G T T - - - - C - - - T - G A!G   DESULFUROCOCCUS MOBILIS
80 ]- U G C]- U[G U G G G G[U]C C]- G C G A -[G G C C C C G C]A -[G C G C C C - - - - C - - - C - A A!G   PYRODICTIUM OCCULTUM
81  - A G U - A[C U G G G G[U]C C]- G C G A -[G G C C C C G G]G - A A A[C C G - - - - C - - - C - G]U[G   OCTOPUS SPRING SPECIES 1
78  - U G U - A[C U G C C C[U]C C]- G G G A -[G G G G G C G G]G - A A G[C C G - G - - G - - - G]- A C[G   THERMOCOCCUS CELER [XTR]
78  - U G U - A[C U G C C C[U]C C]- G A G A -[G G G G G C G G]G - A A -[C C G - G - - G - - - G]- A C[G   PYROCOCCUS WOESEI
75  - U G U - A[C U G U A U[G]C C]- G C G A -[G G G U A C G G]G - A A G[C G C - A - - A - - - U - A]U[G   THERMOPLASMA ACIDOPHILUM
79  - G G T - A[C T G C A A - C]- - G T T A - -[G T T G C G G]G - A A G[C C T - C - - C - - - A - A]C[G   METHANOTHERMUS FERVIDUS
78  U U G U - A[C U G U G G - A C]- G A G A -[G U C U A U G G]G - A A G C U[C - A]- - U - - - A - A C[G   METHANOBACT. FORMICICUM
80  G U G U - A[C U A U G G - G U]- U C C G -[G U C U A U G G]G - A A U U U[C - A]- - U - - - U - U A[G   METHANOBREVIBACTER RUMINANTIUM
80  G U G U - A[C U G C G G - U]U - U U U - - U[G C U G U G G]G - A A G[C C C - A - - C]- - - U - U C[A   MBACT. THERMOAUTOTROPHICUM A1
81  G U G U - A[C U G C G G - U]U - U U U - - U[G C U G U G G]G - A A G[C C C - A - - C]- - - U - U C[A   MBACT. THERMOAUTOTROPHICUM A2
76  G U G U - A[C U G C G G - U]U - U U U - - U[G C U G U G G]G - A A G[C C C - A - - C]- - - U - U C[A   MBACT. THERMOAUTOTROPHICUM B
76  G U G U - A[C U G C G G - U]U - U U U - - U[G C U G U G G]G - A A G[C C C - A - - C]- - - U - U C[A   METHANOBACT. THERMOFORMICICUM
79  - A G U - A[C U G C C]- - - - - A U U U - - - -[G G U G G]G - A A C[A G G - G - - U - - - G]- A C[G   MCOCCUS THERMOLITHOTROPHICUS
79  - A G T - A[C T G C T]- - - - - A T C T - - - -[A G T G G]G - A A C[A A A - G]- - T - - - G - A C[G   METHANOCOCCUS VANNIELII [LNK]
79  - A G T - A[C T G C C]- - - - - A T A C - - - -[G G T G G]G - A A C[A A G - A]- - T - - - G - A C[G   METHANOCOCCUS VOLTAE [LNK]
79  - A G T - A[C T G C T]- - - - - A T C T - - - -[A G T G G]G - A A C[A A G - G]- - T - - - G - A C[G   METHANOCOCCUS VANNIELII [XTR]
79  - A G T - A[C T G C C]- - - - - A T A T - - - -[G G T G G]G - A A C[A A G - A]- - T - - - G - A C[G   METHANOCOCCUS VOLTAE [XTR]
78  - G G U - A[C U G A G G[U]A C]- G C G A -[G U C C U C G G]G - A A A[U C A - U - - C - - - C]- U C[G   METHANOSPIRILLUM HUNGATII
80  - U G U - A[C U A A A G[U]A U]- G A G A -[G U C U U U G G]G - A A A[U A U - G - - G - - - A - A]C[G   METHANOLOBUS TINDARIUS A
77  - U G U - A[C U A A A G[U]A U]- G A G A -[G U C U U U G G]G - A A A[U A U - G - - G - - - A - A]C[G   METHANOLOBUS TINDARIUS B
78  C U G U - A[C U G A A G[U G]U]- G C G A -[G - C U U C G G]G - A A C U C U -[G - - G - - - A]- U C[G   METHANOSARCINA VACUOLATA
79  - U G U - A[C U G A A G[U]G U]- G C G A -[G C C U U C G G]G - A A C U C U -[G - - G - - - A]- U C[G   METHANOSARCINA BARKERI
77  - U G U - A[C U G A A G[U]G U]- G C G A -[G C C U U C G G]G - A A C U C U -[G - - G - - - A]- U C[G   METHANOSARCINA ACETIVORANS
77  - U G U - A[C U G A G G[U]A C]- - A G A -[G U C C U C G G]G - A A A U C U -[G - - C - - - G]- U C[G   METHANOCOCCOIDES METHYLUTENS
77  - G G U - A[C U A A A G[U]G C]- G C G A -[G C C U U U G G]G - A A C[U G C - G - - G - - - A]- U A[G   'METHANOCOCCUS' HALOPHILUS
77  - G G U - A[C U A A A G[U]G C]- G A G A -[G C C U U U G G]G - A A C[U A U - A - - C]- - - U - C C[G   METHANOHALOBIUM EVESTIGATUS
76  - A G U - A[C U G G A G[U]G A]- G C G A -[U C C U C U G G]G - A A C[C A C - G - - A]- - - G - U C[G   HALOBACT.DISTRIBUTUS BKM 4
76  - A G U - A[C U G G A G[U]G G]- G A G A -[C C C U C U G G]G - A G C[G C U - G - - G - - - U]- U C[G   HALOBACT.DISTRIBUTUS BKM13 [U]
76  - A G U - A[C U G G A G[U]G G]- G C G A -[C C C U C U G G]G - A A C[G U C - G - -<A]- - - U - U C[G   HALOBACT.DISTRIBUTUS BKM13 [M]
77  - A G U - A[C U G G A G[U]G A]- G C G A -[U C C U C U G G]G - A A A U[C C - A - - G]- - - U - U C[G   HALOFERAX VOLCANII
78  - A G U - A[C U G G A G[U]G A]- G C G A -[U C C U C U G G]G - A A A[U>C C - A - - G]- - - U - U C[G   HALOFERAX MEDITERRANEI
75  - A G U - A[C U G G A G[U]G C]- G C G A -[G C C U C U G G]G - A G A[G C U - G - - A]- - - U - U C[G   HALOCOCCUS TURKMENICUS
77  - A G U - A[C U G G A G[U]G U]- G C G A -[A C C U C U G G]G - A A A[A>C U ^ G - - G]- - - U - U C[G   HALOCOCCUS MORRHUAE
77  - A G T - A[C T G G A G[T]G C]- G C G A -[G C C T C T G G]G - A A A[T C<C>- G - - G]- - - T - T C[G   HALOARCULA MARIS-MORTUI
77  - A G U - A[C U G G A G[U]G C]- G C G A -[G C C U C U G G]G - A A A U[C C - G - - G]- - - U - U C[G   HALOBACTERIUM SALINARIUM
76  - A G U - A[C U G G A G[U]G G]- G A G A -[C C C U C U G G]G - A G A[G C U - G - - A - - - U]- U C[G   NATRONOCOCCUS OCCULTUS
75  - A G U - A[C U G G A G[U]G G]- G A G A -[C C C U C U G G]G - A G A[G C U - G - - A - - - U]- U C[G   NATRONOBACT. PHARAONIS BKM 6
75  - A G U - A[C U G G A G[U]G G]- G A G A -[C C C U C U G G]G - A G A[G C C - G - - A - - - U]- U C[G   NATRONOBACT. PHARAONIS BKM 12
```

	310 . 320 . 330 .		
113	C U G G G[A]U - G G G u]u u u -	SULFOLOBUS ACIDOCALDARIUS +	(SULFOLOBALES, SULFOLOBACEAE)
113	C T G G G[A]T - G G G C]T T T -	SULFOLOBUS B12	(SULFOLOBALES, SULFOLOBACEAE)
111	C U G G G[A]U - G G]A C A U U -	SULFUROCOCCUS MIRABILIS	(SULFOLOBALES, SULFOLOBACEAE)
119	C T G G G[A]T C G G G C A C C]-	DESULFUROCOCCUS MOBILIS	(TH.PROTEALES, DESULFUROCOCC.)
117	C C G G G[A]U C G G G C C G]- -	PYRODICTIUM OCCULTUM	(SULFOLOBALES, PYRODICTIACEAE)
118	C U G G G - A - G G G G]G C U U	OCTOPUS SPRING SPECIES 1	(TH.PROTEALES, TH.PROTEACEAE ?)
116	C C G C C - G - G - C C]A C u -	THERMOCOCCUS CELER [XTR]	(THERMOCOCCALES)
115	C C G C C - G - G - C C]A - - -	PYROCOCCUS WOESEI	(THERMOCOCCALES)
113	C U G U U -<A>- - - C C]A C u -	THERMOPLASMA ACIDOPHILUM	(THERMOPLASMATALES)
114	T C G C C - G - G - C C C]A - -	**METHANOTHERMUS FERVIDUS**	**(METHBACTERIALES, -THERMACEAE)**
116	C U G C C - G - G - C C U]C - -	**METHANOBACT. FORMICICUM**	**(METHBACTERIALES, -BACTERIACEAE)**
118	C U G C C - A - G - C U U]U U U	**METHANOBREVIBACTER RUMINANTIUM**	**(METHBACTERIALES, -BACTERIACEAE)**
117	C U G C C - A - G - C C]A C u -	**MBACT. THERMOAUTOTROPHICUM A1**	**(METHBACTERIALES, -BACTERIACEAE)**
118	C U G C C - A - G A C C]A C c c	**MBACT. THERMOAUTOTROPHICUM A2**	**(METHBACTERIALES, -BACTERIACEAE)**
113	C U G C C - A - G[A]C]A A - - -	**MBACT. THERMOAUTOTROPHICUM B**	**(METHBACTERIALES, -BACTERIACEAE)**
113	C U G C C - A - G - - C]A C C C	**METHANOBACT. THERMOFORMICICUM**	**(METHBACTERIALES, -BACTERIACEAE)**
110	C U G U U - G - A - U C A]C - -	**MCOCCUS THERMOLITHOTROPHICUS**	**(METHANOCOCCALES)**
110	C C G T T - A - G - T C A]C - -	**METHANOCOCCUS VANNIELII [LNK]**	**(METHANOCOCCALES)**
110	C T G T T - A - G - T C A]C - -	**METHANOCOCCUS VOLTAE [LNK]**	**(METHANOCOCCALES)**
110	C T G C C - G - A - T C A]C - -	**METHANOCOCCUS VANNIELII [XTR]**	**(METHANOCOCCALES)**
110	C T G C C - G - A - T C A]C - -	**METHANOCOCCUS VOLTAE [XTR]**	**(METHANOCOCCALES)**
116	C U G C U - A - U U G]U U - - -	**METHANOSPIRILLUM HUNGATII**	**(METHMICROBIALES, -MICROBIACEAE)**
118	C U G C C - A -<U>C U C C]- - -	**METHANOLOBUS TINDARIUS A**	**(METHMICROBIALES, -SARCINACEAE)**
115	C U G C C - A - - C - C]A A C -	**METHANOLOBUS TINDARIUS B**	**(METHMICROBIALES, -SARCINACEAE)**
116	- U G C C - A<U>A - U C]A G U -	**METHANOSARCINA VACUOLATA**	**(METHMICROBIALES, -SARCINACEAE)**
117	C U G C C - A<U>A C U C]A C C -	**METHANOSARCINA BARKERI**	**(METHMICROBIALES, -SARCINACEAE)**
115	C U G C C - A - A]C U U C - - -	**METHANOSARCINA ACETIVORANS**	**(METHMICROBIALES, -SARCINACEAE)**
114	C U G C U - A - G C]G G X - - -	**METHANOCOCCOIDES METHYLUTENS**	**(METHMICROBIALES, -SARCINACEAE)**
115	C U G C C - A - G]C U C C - - -	**'METHANOCOCCUS' HALOPHILUS**	**(METHMICROBIALES, -SARCINACEAE)**
115	C U G C U - - - G A]U C A A U -	**METHANOHALOBIUM EVESTIGATUS**	**(METHMICROBIALES, -SARCINACEAE)**
114	C C G C C - U - U C]C C - - - -	HALOBACT.DISTRIBUTUS BKM 4	(HALOBACTERIALES)
114	C C G C C - U - A C]U C - - - -	HALOBACT.DISTRIBUTUS BKM13 [U]	(HALOBACTERIALES)
114	C C G C C - U]- X - - - - - - -	HALOBACT.DISTRIBUTUS BKM13 [M]	(HALOBACTERIALES)
115	C C G C C]- - - C C U - - - - -	HALOFERAX VOLCANII	(HALOBACTERIALES)
116	C C G C C - C]- C U - - - - - -	HALOFERAX MEDITERRANEI	(HALOBACTERIALES)
113	C C G U C - - - A]C U - - - - -	HALOCOCCUS TURKMENICUS	(HALOBACTERIALES)
115	C C G C C - U]- C C C - - - - -	HALOCOCCUS MORRHUAE	(HALOBACTERIALES)
115	C C G C C]- - - A C C - - - - -	HALOARCULA MARIS-MORTUI	(HALOBACTERIALES)
115	C C G C C - U]- A C U - - - - -	HALOBACTERIUM SALINARIUM	(HALOBACTERIALES)
114	C U G C U - U - U A]- - - - - -	NATRONOCOCCUS OCCULTUS	(HALOBACTERIALES)
113	C C G C C - U - U]- - - - - - -	NATRONOBACT. PHARAONIS BKM 6	(HALOBACTERIALES)
113	C C G C C - U - U]- - - - - - -	NATRONOBACT. PHARAONIS BKM 12	(HALOBACTERIALES)

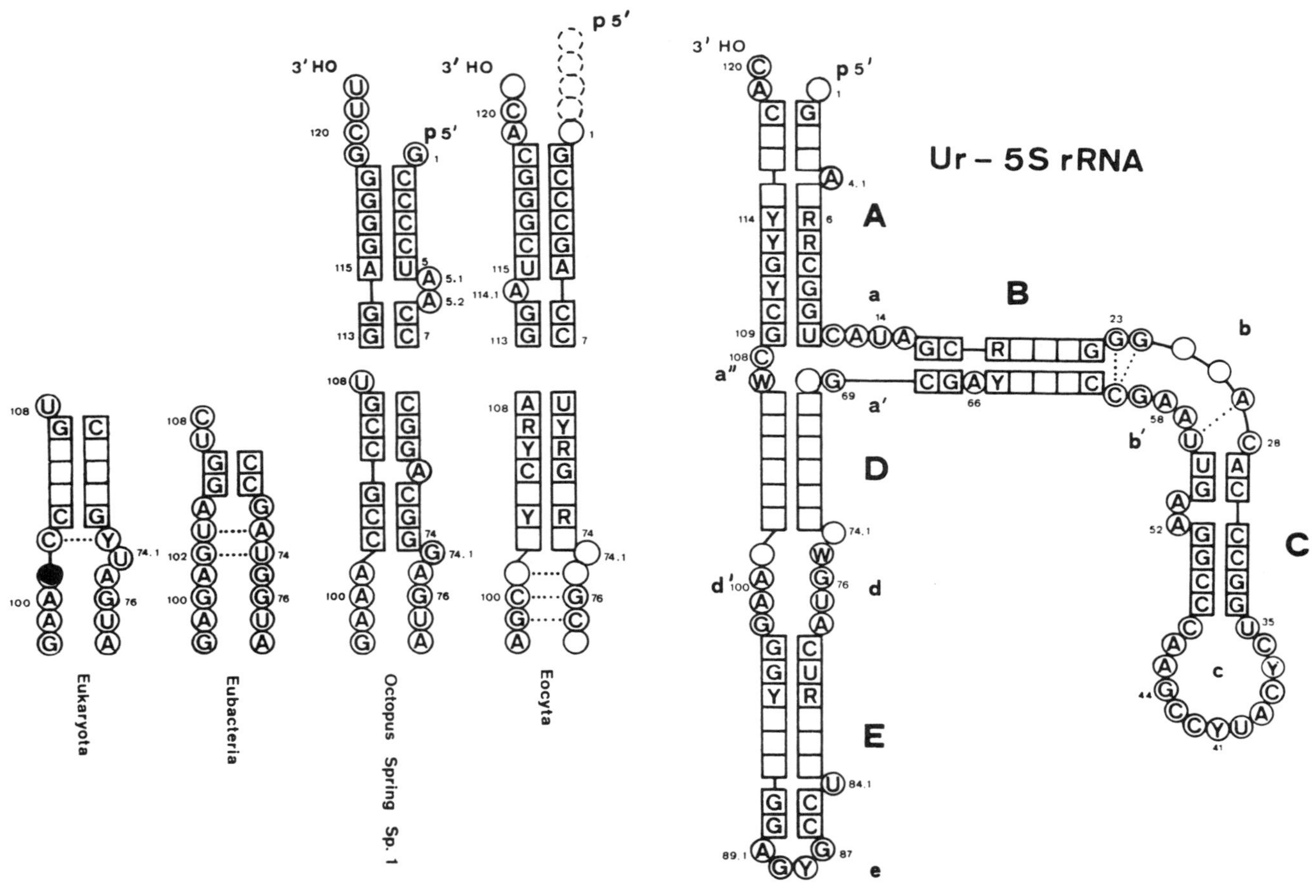

Figure 1 Minimal model of 5S rRNA secondary structure. (*Squares*) Conserved base-pairing; (*circles*) unpaired nucleotides; (*dotted lines*) possible helix extensions. Bases indicated in the minimal model are supposedly plesiomorphic (ancestral) to the respective group, hypervariable positions remain blank. The numeration of bases is according to the *E. coli* sequence (Erdmann 1977). Differences between the five structural groups are shown for helix/loop D/d and Helix A. (Reprinted, with permission of Oxford University Press, from Specht et al. 1990.)

REFERENCES

Cornish-Bowden, A. 1985. Nomenclature for incompletely specified bases in nucleic acid sequences: Recommendation 1984. *Nucleic Acids Res.* **13:** 3021–3030.

Erdmann, V.A. 1977. Collection of published 5S and 5.8S ribosomal RNA sequences. *Nucleic Acids Res.* (suppl.) **5:** r1–r13.

Specht, T., V.A. Erdmann, and J. Wolters. 1991a. The Berlin RNA Databank: Compilation of 5S rRNA and 5S rRNA gene sequences. In *Protocols for Archaebacterial Research* (ed. E.M. Fleischmann et al.), pp. D1–D3. Center of Marine Biotechnology Press, Baltimore.

——. 1991b. Compilation of 5S rRNA and 5S rRNA gene sequences. *Nucleic Acids Res.* (suppl.) **19:** 2189–2191.

Specht, T., J. Wolters, and V.A. Erdmann. 1990. Compilation of 5S rRNA and 5S rRNA gene sequences. *Nucleic Acids Res.* (suppl.) **18:** 2215–2230.

Appendix 6

16S and 23S rRNA-like Primers

Compiled by L. Achenbach and C. Woese

Small- and large-subunit rRNA sequences have long been standard for phylogenetic analysis. There are currently 4456 molecules of 16S–18S size and 891 molecules of 22S–28S size in the ribosomal database, which also includes secondary structure and alignment files. The Ribosomal Database Project (RDP) can be accessed via ftp (rdp.life.uiuc.edu: pub/RDP) or gopher (rdpgopher.life.uiuc.edu). The primer list that follows is one that has accumulated over the years and is fairly comprehensive for those primers that work well on Archaea. Because these primers have been used almost exclusively for sequencing, only a few have been tried in PCR amplifications. Typically, terminal primers (i.e., 25e Forward and 1525 Reverse) are used to amplify the rDNA, and internal primers are used for direct sequencing of the amplification product. Please note that primer locations are based on *Escherichia coli* positions and that IUPAC notation has been used throughout.

16S and 23S rRNA-like Primers

Primer name	*E. coli* numbers	Sequence	Archaea	Bacteria	Eukarya
16S-like Primers					
0008 Reverse	25–08	CTGGCAGGATCAACCGGA	±		
0011 Reverse	28–11	CCTGAGCCAGGATCAAAC	±		
0023a Forward	7–23	CTCCGGTTGATCCTGCC	±		
0025e Forward	9–25	CTGGTTGATCCTGCCAG	++		+
0112a Reverse	128–112	CCACGTGTTACT(S)AGC	++		
0261 Forward	246–261	AGCTAGTTGGTGGGGT	+	+	
0342 Reverse	357–342	CTGCTGC(S)(Y)CCCGTAG	±	+	
0343as Reverse	357–343	CTGCTGCGCCCCGTA	+ (for crenarchaes)		
0348a Forward	333–348	TCCAGGCCCTACGGG	++		
0498ca Reverse	519–498	ACACCAG(R)CTTGCCCCCCGCTT	for crenarchaes		
0498ea Reverse	511–498	CTTGCCC(R)GCCCTT	for eukaryarchaes		
0517 Reverse	531–517	ACCGCGGC(K)GCTGGC	+	++	++
0519 Reverse	536–519	GWATTACCGCGGCKGCTG	+	+	
0530 Forward	515–530	GTGCCAGC(M)GCCGCGG	±	+	
0690a Reverse	704–690	TTACAGGATTTCACT	++		
0802 Forward	787–802	ATTAGATACCCTGGTA	+	+ (taxa-specific)	
0802a Forward	787–802	ATTAGATACCCGGGTA	+	+ (taxa-specific)	
0915a Reverse	934–915	GTGCTCCCCCGCCAATTCCT	+		
0922 Forward	906–922	GAAACTTAAA(K)GAATTG	+		
0956 Reverse	973–956	GGCGTTGTGTC(S)AATTAA	+	+	
1068 Forward	1053–1068	GCATGGC(Y)G(Y)CGTCAG	++	+	
1100 Reverse	1115–1100	AGGGTTGCGCTCGTTG	±	++	+
1100a Reverse	1115–1100	TGGGTCTCGCTCGTTG	++		
1114 Forward	1099–1114	GCAACGAGCGCAACCC	+	++	
1226a Reverse	1242–1226	CCATTGTAGC(S)CGCGTG	+		
1240 Forward	1225–1240	ACACGCGTGCTACAAT	+	+ (taxa-specific)	

1388a Reverse	1407–1388	GACGGGCGGTGTGTGCAAGG	+		
1392 Reverse	1406–1392	ACGGGCGGTGTGT(R)C	++	++	++
1406 Forward	1391–1406	TG(Y)ACACACCGCCCGT	++	++	
1492 Reverse	1510–1492	GGTTACCTTGTTACGACTT	+	+	+
1525 Reverse	1541–1525	AAGGAGGTGATCCAGCC	++	++	
1538 Forward	1524–1538	CGGTTGGATCACCTC	++		
23S-like Primers					
0189 Reverse	207–189	TACTTAGATGTTTCAGTTC			
0213a Reverse	228–213	GTTGGTTTCTTTTCCT	++	+	
0227a Forward	212–227	GAGGAAAAGAAATCAA	±	–	
0339 Forward	322–339	AGAGGGTGAAAGCCCCGT	+		
0420 Reverse	436–420	GACGTATTTAGCCTTGG	+		
0459 Reverse	473–459	CTTTCCCTCACGGTA	++	++	
0481 Forward	464–481	TGAGGGAAAGGTGAAAAG	++		
0567a Reverse	582–567	CGGCCCTTGTTTCAAG	+	–	
0775 Reverse	790–775	AT(K)GG(Y)CTTTC(R)CCCC	+	++	+
0821 Forward	806–821	CTGGTTCTCCCCGAAA	++	++	
1075 Forward	1061–1075	TGGCTTAGAAGCAGC	+	+	
1600 Reverse	1617–1600	GTGTCGGTTTTGGGTACG	±	+	
1941 Reverse	1956–1941	AACTTACCCGGCAAGG	++	++	–
2576a Reverse	2591–2576	GGTCTAAACCCAGCTC	++	–	
2586 Forward	2572–2586	GCGTGAGCTGGGTTT	++	++	
2759 Forward	2744–2759	GCTGAAAGCATCTAAG	++	++	
2840a Reverse	2855–2840	TGCGTACACCCCGAG	++		

– = Poor (does not work).
± = May work on some.
+ = Fair to good.
++ = Very good.

Appendix 7

Thermodynamic Data on the Activation and Denaturation of Proteins from Thermophilic Archaea

Contributed by H.H. Klump

Many proteins from thermophiles and hyperthermophiles display extraordinary thermostability. Much of our information concerning the structural transitions and especially the complete thermal unfolding of proteins has come from scanning calorimetry. A very simplified description of an adiabatic differential scanning microcalorimeter must focus on the following components: Samples of a dilute aqueous protein solution containing approximately 2–4 mg of protein dissolved in 1 ml of a suitable buffer and a buffer blank are filled into pressurizable sample and reference cells, respectively, and heated in parallel. The difference in energy input required to keep the samples at the same temperature is precisely monitored. As the transition temperature (T_m) is reached, an additional energy input into the sample cell is required to compensate for the energy consumed by the cooperative unfolding of the protein. The total excess heat capacity change due to the denaturation generates a peak over the instrument base line. Comparing the peak area to an electrically induced calibration area allows calculation of the energy of unfolding of the polymer. The experiment reveals the transition temperature, the degree of cooperativity as reflected in the sharpness of the peak, and the total energy ΔH, entropy ΔS, and free-energy ΔG required. A moderate pressure of 4 kg/cm^2 shifts the boiling temperature of the solvent water away from the temperature of unfolding. High-sensitivity DSC (HSDSC) up to 130°C, which requires pressurable cells, can only be performed with the DASM-4 and the most recent version DASM-4M (see Commentary). Comparable instruments for less demanding tasks in terms of temperature range available are the MC-2 from Microcal (Amherst, Massachusetts), which has a cut-off temperature of 110°C, and the Bio DSC from Sceres Instruments (France). Measurements on proteins from hyperthermophiles are out of reach for all other commercially available instruments. The MC-2 and the Bio DSC from Sceres Instruments exhibit characteristics similar to those of DASM-4 in terms of sample size or signal-to-noise ratios, effective cell volume, or relative error in heat capacity determinate. Not quite the same performance can be expected from the medium-sensitivity DSC instruments such as Perkin Elmer DSC-7, although they are run on biological samples. Since the restructuring of the USSR, it is expected that the DASM-4 will be more readily available in the near future.

The instrument proper consists of two units: the sample compartment and the control unit. It is interfaced with a PC and operated by the keyboard through an interactive program. To chose the adequate experimental conditions (i.e., sample concentration, buffer, heating rate, cut-off temperature, and solvents to clean the capillary cells after thermal denaturation and possible aggregation of the protein, etc.), it is advisable to run a melting experiment under comparable conditions in a cuvette using a spectrophotometer capable of showing a characteristic change in an optical signal due to denaturation. A Hewlett-Packard Diode Array Spectrophotometer with a thermostatted cell compartment has this capability.

How To Set Up a Calorimetric Scan

Sample Preparation

To run the instrument at its highest sensitivity setting, it is very important that the heat capacity of the sample matches that of the reference solution as closely as possible. It is thus advisable to dialyze the protein for 1 day further against the appropriate buffer solution and use the dialysis medium as the reference solution. If the experiment requires excessively high concentrations of protein, it may be advisable to add an appropriate amount of amino acids to the reference solution as well. For this reason, the use of an appropriate buffer such as glycylglycine buffer may be helpful.

Filling Procedure

Prior to filling the two capillary cells with the sample and the reference solution, degas the liquids by keeping them in an evacuated desiccator for 10 minutes. Then, fill a modified Eppendorf pipette carefully with the buffer so that no gas bubbles are sucked up. Load the solution into the reference cell in such a way that the capillary is completely filled. To avoid air from entering into the cell, close one side of the capillary with a fitted Teflon stopper. Load the sample solution into the other coiled platinum capillary accordingly and insert a stopper at one side. Close the sample compartment with a screw cap and apply external gas pressure to the open side of each capillary. In most cases, 2.5 atm pressure is sufficient. It is important to fill the sample into the sample cell and the reference solution into the reference cell to enable the electronic controllers to function properly. Only if the expected result is an exothermic reaction (as in the case of aggregation or self-assembly mechanisms) is it advisable to switch the assignment of the cells. In case of a base line scan, fill the two capillaries with water or with the same buffer as used for the reference.

Scanning

After the cells have been properly filled and pressurized, the computer will provide advice on how to proceed with the correct key strokes to achieve a successful scan. All necessary routines to compensate for the energy required for the thermal unfolding of the dissolved protein are done by the instrument's electronics. The actual status during a scan can be watched on the screen. The first scan of the series of scans should always be a buffer/buffer scan to check for the performance of the instrument. For the second scan, the content of the sample cell should be exchanged for the sample solution. At any position outside the temperature interval of the thermal transition, an electronic calibration peak can be set. This usually represents an area equivalent to 2.8 mcal when the power of 50 x 10^{-6} VA is applied for 240 seconds. After the completion of a scan, the data are stored and can be called up for further processing such as subtracting the buffer base line. All necessary calculations are included in the software package. The final results can be plotted as graphs or printed as tables. The time required for completion of a scan is approximately 2 hours.

Data Analysis

DSC data interpretation is normally based on the equilibrium thermodynamic expression: $[d\ln K/dT]_p = \Delta H_{vH}/RT^2$, where K is the equilibrium constant of the process, T is the absolute temperature, and ΔH_{vH} is van't Hoff enthalpy. R denotes the gas constant, and the index P states that the equation holds at constant pressure. This equation is directly applicable only to two-state processes, i.e., reactions in which intermediate

Table 1 Typical Thermodynamic Data from Proteins from *Pyrococcus furiosus*

Denaturation	ΔH_{cal}	ΔS_{cal}	T_m
Rubredoxin (oxidized)	22.5	57.0	113
Rubredoxin (reduced)	20.2	53.8	102
Rubredoxin (zinc-conjugated)	23.8	60.0	123
Ferredoxin (oxidized/4 Fe)	11.5	29.3	117.5
Ferredoxin (reduced/4 Fe)	11.4	29.0	117.0
Ferredoxin (oxidized/3 Fe)	11.0	29.3	105
Glutamate dehydrogenase	414	1072	113

Data from Klump et al. (1994).

states between the initial (native) and final (denatured) states are not significantly populated at equilibrium. For the two-state process, T_m should be independent of the concentration. The following van't Hoff expression also applies for two-state processes: $\Delta H_{vH} = ART^2\ C_{ex}/\Delta h_{cal}$, where Δh_{cal} is the calorimetric specific enthalpy, C_{ex} is the apparent excess heat capacity at T_m, and A is a constant, which for a two-state process has the value of 4.0. This enthalpy ΔH_{cal} is a measure of the total heat involved in the process. One other mathematical expression is useful for a strictly two-state process carried out under essentially equilibrium conditions: $\Delta H_{vH} = \Delta H_{cal} = \Delta h_{cal}\ M$, where M is the molecular weight of the molecule under investigation. If $\Delta H_{cal} < \Delta H_{vH}$, then intermolecular cooperation is clearly indicated. This treatment is strictly valid only for reversible processes that are subject solely to thermodynamic limitations. Under these conditions, $\Delta H_{cal}/T_m = \Delta S_{cal}$, and finally, under the assumption that ΔH is temperature-independent, the free enthalpy at any temperature T becomes $\Delta G(T) = \Delta H\,(1 - T/T_m)$.

Table 1 gives representative data on the values for ΔH_{cal}, ΔS_{cal}, and T_m for the proteins from the hyperthermophile, *Pyrococcus furiosus*.

Instrument Specifications for the DASM-4

Electronic noise (heating rate: 1°C/minute)	5×10^{-7} W
Utilizable temperature range	0–130°C
Thermal noise (heating rate: 1°C/minute)	15×10^{-6} W
Reproducible base line	3×10^{-6} W
Heating rate °C/minute	0.125, 0.25, 0.5, 1.0, 2.0
Cell volume milliliter	0.6
Cooling by water (for peltiers)	
Energy uptake VA	300

COMMENTARY

- *The National Center for Biocalorimetry:* DASM-4M is housed in the United States and can be accessed at the National Center of Biocalorimetry. Before planning to use the instrument at the National Center, contract Professor E. Freire in the Department of Biology, Johns Hopkins University, Baltimore, Maryland, to arrange for a time slot and for assistance. The instrument was developed and manufactured by a special workshop for biocalorimetry at the Institute of Protein Research of the Soviet Academy of Science, Poushino, Moscow Area, USSR, and was marketed through Mashpriborintorg.

• *Important Precautions:* The advantages of the spectroscopic experiment prior to the calorimetric scan is twofold: (1) It allows for exploration of a set of experimental conditions for the investigation of a protein, such as pH, concentration of protein, buffer, reversibility, and heating rate; the temperature interval in the heat-induced cooperative conformational change can be observed with about tenfold less material as one would need for a calorimetric experiment. (2) Possibly even more important, it avoids the embarrassing confession that a protein has precipitated and clogged the capillary cells of the DASM-4M.

• *Applications:* The DASM-4M is a research-grade instrument that records changes in excess heat capacity of dissolved biopolymers due to conformational changes. The range of application includes molecular biology, biophysics, physical chemistry, and polymer science. The following problems can be tackled with the aid of microcalorimetry: (1) interactions between proteins and lipids; (2) assembly processes in proteins; (3) conformational changes of synthetic polymers in organic solvents; (4) analysis of equilibrium states; (5) identification of plasmids in solution; and (6) identification of nucleic acids and virions.

REFERENCES

DiRuggiero, J., F.T. Robb, R. Jagus, H. Klump, K.M. Bores, M. Kessel, X. Mai, and M. Adams. 1993. Characterization, cloning and in vitro expression of the extremely thermostable glutamate dehydrogenase from the hyperthermophilic archaeon, ES4. *J. Biol. Chem.* **268:** 17767–17774.

Klump, H.H. 1988. Conformational transitions in nucleic acids. In *Biochemical thermodynamics*, 2nd edition (ed. M.N. Jones), pp. 100–144. Elsevier Science, Amsterdam.

Klump, H.H., M.W. Adams, and F.T. Robb. 1994. Life in the pressure cooker. *Pure Appl. Chem.* **66:** 485–489.

Klump, H.H., J. Di Ruggiero, M. Kessel, J.-B. Park, M.W.W. Adams, and F.T. Robb. 1992. Glutamate dehydrogenase from the *Hyperthermophile Pyrococcus furiosus. J. Biol. Chem.* **267:** 22681–22685.

Privalov, P.L. and S.A. Potekhin. 1988. Scanning microcalorimetry in studing temperature-induced changes in proteins. *Methods Enzymol.* **131:** 4–51.

Appendix 8

Suppliers*

Altair, 1171 Ocean Avenue, Oakland, CA 94608. Telephone 510-547-8337. Fax 510-450-9748.

BBL Microbiology Systems, P.O. Box 243, Cockeysville, MD 21030. Telephone 410-771-0100 or 800-638-8663.

Bioblock Scientific, Parc d'innovation, Starsbourg-Illkirch, BP 111, F-67403 IllKirch cedex, France. Telephone 33-88-67-14-14. Fax 33-88-67-11-68.

Büchi Labortechnik AK, Postfach Flawil 9230, Switzerland. Telephone 41-71-84-6363. Fax 41-71-83-5711.

Chiswick Trading Inc., 33 Union Ave., Sudbury, MA 01776-2267. Telephone 508-443-9592. Fax 1-800-638-9899.

Don Whitley Scientific Ltd., Green Lane, Bardow, Shipley, W. Yorks, England. Distributed in France by: AES Laboratoire, route de Dol, B.P. 27, F-35270 Combourg, France. Telephone 33-99-73-11-55

General Oceanics, Inc., 5535 N.W. 7th Avenue, Miami, FL 33127. Telephone 305-754-6658.

Glas-Col Inc., 711 Hulman Street, Terre Haute, IN 47802. Telephone 812-235-6167. Fax 812-234-6975.

Haskel, 100 East Graham, Burbank, CA 91502. Telephone 818-843-4000. Fax 818-843-4375.

High Pressure Equipment Co., 1224 Linden Avenue, Erie, PA 16505. Telephone 814-838-2028. Fax 814-838-6075

Independent Equipment Corp. (IEC), 5 Johnson Dr., P.O. Box 130, Raritan, NJ 08869-0130. Telephone 908-526-1001. Fax 908-526-7887.

La Calhène S.A., 1 rue du petit Clamart, F-78140 Vélizy-Villacoublay cedex, France. Telephone 33-01-46-30-66-00. Fax 33-01-46-30-8730.

Markson Science Inc., 10201 S. 51st Street, Phoenix, AZ 85044. Telephone 800-528-5114. Fax 800-858-2243.

Merck & Co., Kelco Division, 500 West Madison Avenue, Suite 31800, Chicago, IL 60661. Telephone 800-535-2687. Fax 312-831-3704.

Microcal, Inc., 22 Industrial Drive East, Northampton, MA 01060. Telephone 413-586-7720. Fax 413-586-0149.

Omega Engineering, Inc., 1 Omega Drive, P.O. Box 4047, Stamford, CT 06907. Telephone 203-359-1660. Fax 203-359-7700.

OSI, 141 rue de Javel, F-75739 Paris cedex 15, France. Telephone 33-01-45-54-97-31. Fax 33-01-45-54-26-28

PolyLabo Paul Block & Cie, 305, route de Colmar B.P. 36, F-67023 Strasbourg cedex, France. Telephone 33-88-65-80-20. Fax 33-88-39-74-41.

Sceres Instruments, 16, rue de Chartes, F-91400 Orsey, France.

SchweizerHall Inc., 10 Corporate Place South, Piscataway, NJ 08854. Telephone 908-981-8200. Fax 908-981-8282.

Thermolyne, 2555 Kerper Blvd., P.O. Box 797, Dubuque, IA 52004-0797. Telephone 319-556-2241. Fax 319-556-0695.

*Additional suppliers not listed in *The 1995 Lab Manual Source Book.*

W.B. Moore, Inc., P.O. Box 1936, Easton, PA 18044. Telephone 610-559-9474 or 800-379-8181. Fax 610-559-1546.

Yellow Springs Instrument Co., P.O. Box 279, Yellow Springs, OH 45387. Telephone 800-765-4974. Fax 513-767-9353.

THERMOPHILES: INDEX

Acetamide, as denaturant, 101, 152
Acetate, as organic substrate in hyperthermophile culture medium, 10
Acidianus, 4
  brierleyi, 175
  infernus medium, 180
  optimal growth conditions, 19
Acidophiles, growth media and preparation, 19–20
Acridine orange stain, 13
Agar
  media
    evaporation at high temperature, 18
    Pyrococcus furiosus growth, 18
  plates, streaking hyperthermophile cell suspensions, 11
Aldehyde ferredoxin oxidoreductase, 67
ALVIN, deep-sea isolation of heterotrophic hyperthermophiles using, 9–10
Amino acid
  determination of essential, hyperthermophile growth and, 12–13
  requirements for heterotrophic hyperthermophilic Archaea, 12–13
  stock solution, hyperthermophile cultivation and, 12
AN1 medium, 180
Anoxic
  hood, 11
  hot water samples from deep-sea hydrothermal vents, 9–10
    heterotrophic hyperthermophile samples, storage, and transportation in Hungate tubes, 11
Antifoam A, 56
Archaeal introns. *See* Introns, archaeal
Archaeoglobus
  commercially available strains, 159
  medium, 183–184
  profundus medium, 185
Artificial sea water, 10, 34
  containing Wolfe's trace minerals, 12
Autoclaving vs. filter-sterilizing culture media, 10
Autotrophic growth of thermophiles and
  H_2 as electron donor, 4
  O_2 as terminal electron acceptor, 4
Autotrophs, 135
  growth, 4
    and S^0-dependent Archaea, 4
Auxotrophic mutants, isolation by replica-plating, 128–131

Balch tubes (anaerobic), 17
  inoculating hyperthermophile samples, 16
Berlin RNA Databank, compilation of 5S rRNA and 5S rRNA gene sequences, 195–200
BS-medium, 185–186
Buffers
  Assay (10x), detection of DNA cleavage activity, 114, 116
  *Bam*HI, 143
  Detergent, hydrogenase solubilization, 63–65
  *Eco*RI, 143
  Elution, membrane preparation and hydrogenase purification, 64–65
  EPPS, 68
  Extraction, 97
  formamide loading, 110, 115
  HEPES/EDTA, 119
  *Hha*I, 150
  *Hpa*II, 150
  Hybridization
    (2x), ribosomal gene removal from total DNA, 120
    (4x), typing marine vent thermophiles, 143
  lysis
    Archaeal introns in rRNA genes, 112
    typing marine vent thermophiles by DNA polymerase RFLP, 140
  mutagenesis, isolation of *Sulfolobus acidocaldarius* mutants, 129
  phosphate, 74
  Reaction (10x), PCR amplification of 16S rRNA genes, 101
  REV (25x), 110
  RNA extraction, 97
  RT (5x), 120
  SDS sample, 115
  SET hybridization wash buffer (0.1x), 143
  Southern transfer, 143
  Splice (10x), 112
  Stop, detection of DNA cleavage activity, 115
  Sulfolobus dilution, 128
  T7 RNA polymerase (10x), 112
  Taq, 150
  TBE, 110
  TE, 95, 97, 115, 140, 149
    10x, 101
  TKE (10x), 110
  TM, 112
  urea loading, 110
  wash, membrane preparation and hydrogenase purification, 63–65
"Bulge-helix-bulge" structure, 107
Butyl septum-type rubber stoppers, 13, 16–17

^{13}C-NMR
  spectrometry, study of glucose metabolic pathways in *Thermoplasma acidophilum*, 81–85
  spectra, 84
Casamino acids, amino acid composition, 12
cDNA cloning, 99
Chain termination, 109
CHAPS, as detergent, extraction of *Pyrodictium abyssi* hydrogenases, 65
Clostridium spp., optimal growth temperatures, 18
Codon usage tables for thermophilic Archaea, 191–194
Coenzyme M, component of nonmarine methanogen culture medium, 21
Colloidal sulfur. *See* Sulfur, colloidal

Column chromatography, protein separation using. *See also* FPLC
hydroxyapatite, hydrogenase/ferredoxin purification, 69–70
Q-Sepharose, hydrogenase/ferredoxin separation, 69
Sepharose CL-6, ferredoxin purification, 70
silicic acid, fractionation of total lipid using, 76–77
Superose 6 HR 16/50, hydrogenase purification, 70
Continuous culture techniques
apparatus, 32
colloidal sulfur preparation for continuous feed, 33
maintaining continuous flow rate in reactor system, 32
reactor system, 31–32
steel vs. Teflon-coated parts, thermoacidophile cultivation, 35
thermoacidophile cultivation using, 35
thermoanaerobe cultivation using, 34–35
Continuous-flow centrifuge vs. Millipore tangential-flow filtration cell, harvesting cells using, 56
Cysteine
vs. elemental sulfur in hyperthermophile culture medium, 10
hydrochloride, 21
-sulfide, as reducing agent, 20
Cytochromes, 63

DEAE-Sepharose Fast Flow FPLC column, hydrogenase/ferredoxin purification using, 69–70
Deep-sea
isolation of heterotrophic hyperthermophile samples
anoxic hot water, 9–10
hot sediments, 10
solid polymetal sulfides, 10
vents, as source of thermophilic Archaea, 4, 9, 15
Desulfobacter
postgatei, 176–177
sp., medium, 177
Desulfurella medium, 184
Desulfurococcus
amylolyticus medium, 182–183
commercially available strains, 159–160
medium, 176
Desulfurolobus, commercially available strains, 160
Desulfuromonas (hyperthermophile) strain SY, 12
Detergent. *See also* CHAPS; Nonidet P-40; SDS
solubilization of hydrogenases, 63–65
Dilution-to-extinction series, obtaining pure hyperthermophile cultures and, 16–17
DNA
biotinylated, 121
cDNA, creation from population of differentially expressed genes, 121–123
chromosomal, isolation, 140
cleavage activity detection, 114, 116
endonuclease activity, rRNA introns from hyperthermophiles and, 114
polymerase RFLPs, 139
analysis, 142–147
quantitation of agarose gels, 142
Double-quantum-filtered correlation spectroscopy, 83
Drierite desiccant, 11

East Pacific Ridge, hydrothermal vent and isolation of heterotrophic hyperthermophiles, 9
Edmond-Walden sampler, hydrothermal fluid collection and
vs. nonsterile samplers, 9
preparation before use, 9
Elemental sulfur. *See* Sulfur, elemental
Endonuclease. *See also* DNA endonuclease activity
cleavage site mapping, 116
site-specific, 114
Entner-Doudoroff pathway, 85
Epifluorescence microscopy, counting hyperthermophile cells using, 13, 16
ES4 cultivation, 96. *See also Pyrococcus endeavori*
high-pressure–high-temperature, 37, 40–42
medium, 173
using continuous culture techniques, 31–35
Ester-linked polar lipid fatty acids, 77
Ether-linked isoprenoid lipids
extraction, fractionation, and derivization scheme, 73
fractionation of total lipid using silicic acid column chromatography, 76–77
glassware preparation for extraction, 75
liberating ether lipids, 78
lipid-extracted residue separation, 76
lyophilization, 75
mild alkaline methanolysis, 77
phase separation, 76
removing water from total lipid sample during phase separation, 76
sample storage, 75
silicic acid column chromatography, 76–77
single-phase extraction, 75–76
storage of solvent-free total lipid, 76
Extremely thermophilic *Pyrodictium* spp., hydrogenase purification, 63–65

Fermentor, large-scale growth of hyperthermophiles using, 47–48
Ferredoxin, 6
assay, 68
detection by visible absorption, 68
as electron donor to hydrogenase, 67
purification from *Pyrococcus furiosus*, 70
calculating concentration, 68
estimating purity by visible absorption, 68
storage of purified ferredoxin, 70
yield after purification, 71
separation of hydrogenase and ferredoxin, 69
Filter-sterilizing vs. autoclaving culture media, 10
5-Fluoro-orotate, 133–135
Formate, as carbon source for marine methanogens, 20
Forward mutation of *Sulfolobus acidocaldarius*, measurement using modified accumulation method (A_0), 133–135
FPLC, hydrogenase/ferredoxin separation and purification. *See also* Column chromatography
DEAE-Sepharose Fast Flow, 69–70
Mono Q HR 5/5, 69–70

Gas chromatograph
automatic sampling and high-pressure–high-temperature thermophile cultivation, 42
monitoring cell growth and metabolism using, 40
Gellan gum. *See* Gelrite
Gelrite, 92
determining concentration for solid media, 19, 27
incubation and streaking hyperthermophiles at high temperature, 11
neutrophile growth and, 19
plaque assay medium containing, 19
plate preparation for *Sulfolobus acidocaldarius* cultivation, 126
polymerization and Na^+ concentration, 27–28
solid media preparation using, 19
solution for supporting and overlay gels, 26
Genomic DNA, preparation from sulfur-dependent hyperthermophilic Archaea, 95–96
Glucose metabolic pathways in *Thermoplasma acidophilum* using ^{13}C- and ^{15}N-NMR

^{13}C-glucose and ^{15}N-ammonium chloride, metabolism in cell-free lysates, 81
identification of metabolites, 83–85
NMR spectrometry
of cell lysates, 82
of freeze-dried lysates, 83
parameters, 83
Glutamate dehydrogenase, 67
Glycolipid, total lipid fractionation and silicic acid column chromatography, 76–77
Graham condenser, 31
Guanidinium isothiocyanate, total RNA preparation and, 108
Guaymas Basin
hot springs, 18
hydrothermal vent, and isolation of heterotrophic hyperthermophiles, 9

H_2 (molecular hydrogen)
activation, catalyzed by hydrogenase, 67
evolution and hydrogenase assay using redox dye, 63
oxidation via hydrogenase, 63
production by *Pyrococcus furiosus*, 67
-uptake hydrogenases, 63
H_2S
accumulation, as indicator of cell growth, 54
colorimetric assay, 54
production
in high-salt-containing media, 6
from *Thermoplasma acidophilum*, 53
S^o reduction to H_2S, 4, 48
toxicity to humans and cultures, 53
Heteronuclear multiple bond coherence (HMBC), 83, 85
Heterotrophic hyperthermophiles. *See* Hyperthermophiles, heterotrophic
High-pressure–high-temperature reactor for thermophilic Archaea cultivation
disassembly and sterilization of system, 44
parts, 37–38
pressurization
hydrostatic, 41–42
hyperbaric, 37, 40–41
preventing salt/sulfide buildup in transfer lines, 39
Pyrex reactor insert, 41
removing samples from reactor without decompression, 44
schematic, 38
stainless steel vessel, effect on cell growth/metabolism, 39–40
Hot sediment hyperthermophile samples, 15
from deep-sea hydrothermal vents, 10
Hungate
technique, 11
isolation and purification of hyperthermophiles using, 11
tubes, 9–10
gas-tight, 11
storage and transportation of hyperthermophile samples, 11
Hydrogenase, 6
ferredoxin as electron donor, 67–68
H_2-uptake, 63
half-life of pure enzyme, 71
purification
from *Pyrococcus furiosus*
analysis by native electrophoresis, 70
assay using methyl viologen, 68
storage of purified enzyme, 70
from *Pyrodictium* spp., membrane preparation and detergent solubilization, 63–65
storage of purified and/or solubilized enzyme, 65
Hydrothermal fluid collection, using Edmond-Walden sampler, 9
Hydroxyapatite column chromatography, hydrogenase/ferredoxin purification, 69–70
Hyperthermophilic Archaea (hyperthermophiles). *See also* Acidophiles; Methanogens; Neutrophiles
Archaeal introns in rRNA genes, 107–116
continuous culture techniques, 31–35
culture and growth on solid medium, 18
definition, 15
doubling times, 21–22
genomic DNA preparation from sulfur-dependent, 95–96
heterotrophic
deep-sea isolation and cultivation
amino acid requirements, 12–13
determining cell densities, 13
growing pure cultures, 11
hydrostatic pressure, effect on cell growth, 13
isolation and purification, 11
media and solutions, 10
nonsalt requiring, culture, 18
sample transportation, storage, and incubation, 11
stationary phases and cell lysis, 48
inoculation, 16
isolation from extremely acidic environments, 16
isolation, growth, and maintenance
growth media and preparation
acidophiles, 19–20
methanogens, 20–21
neutrophiles, 17–19
incubation, 21
isolation and maintenance, 16–17
minimum growth temperatures, 18
sampling methods, 15–16
large-scale growth, 47–48
lysis at high temperatures, 22
obtaining pure cultures using dilution-to-extinction series, 16
16S rRNA gene amplification by PCR, 101–105
sulfur-dependent, genomic DNA preparation, 95–96
Hyperthermus
butylicus medium, 184
commercially available strains, 160

Incubation of heterotrophic hyperthermophiles, 11
Introns, archaeal
approximate locations of oligonucleotide primers in relation to cleavage sites, 113
identifying cleavage sites, 113
intron-exxon boundaries, 107
in rRNA genes of hyperthermophiles
detection of in vivo splicing, 109–111
in vitro splicing of transcript from cloned intron-containing genes, 111–113
proteins encoded by rRNA introns from hyperthermophiles, 114–116
two-dimensional PAGE of total RNA, 108–109
In vitro splicing of transcript from cloned intron-containing rRNA genes, 111–113

Lipids. *See* Ether-linked isoprenoid lipids; Glycolipid; Neutral lipid; Polar lipid
N-Lauroyl sarcosine, 87–88

Magic methanol, 74
ether-lipid isolation using, 78
hydroxydiether lipids, degradation by, 79
Maltose, 34
as carbon source, 48
as organic substrate in hyperthermophile culture medium, 10
Marine
broth 2216, 18
cultivation of heterotrophic hyperthermophiles, 10
geothermal vents, 15
vs. springs and geysers, effect of temperature on thermophiles, 6
hypothermophilic methanogens, medium, 20–21

Marine (*continued*)
vent thermophiles
chromosomal DNA isolation, 140–142
growth, 139–140
RFLP analysis, 142–143
DNA digestion, 144
RFLP hybridization, 145
Southern blot transfer, 144–146
Media
Acidianus
brierleyi, 175
infernus, 180
acidophile growth medium, 19–20
AN1, 180
Archaeoglobus, 183–184
profundus, 185
basal salts, for *Pyrococcus furiosus*, 169–170
Brock's, 156
supplemented with carbon sources, 91
BS, 185–186
cultivation of heterotrophic hyperthermophiles
filter-sterilization, 10
gassing with oxygen-free N_2, 10–11
culture and dilution, for sulfur-metabolizing Archaea, 25–26
Desulfobacter
postgatei, 176–177
sp., 177
Desulfurella, 184
Desulfurococcus, 176
amylolyticus, 182–183
enrichment, 168–169
ES4 cultivation, 173
growth
liquid, for *Sulfolobus*, 172
for *Methanococcus jannaschii*, 170–171
for *Thermoplasma acidophilum*
7A, 171–172
7B, 172
heterotrophic hyperthermophilic Archaea, 167–168
Hyperthermus butylicus, 184
large-scale
culture of *Thermoplasma acidophilum*, 55–56
growth medium for *Pyrococcus furiosus*, 171
methanogen growth medium, 20–21
Methanogenium, 173–174
Methanothrix, 181
MG, 185
MTP4, 185
neutrophile growth medium, 17–19
plaque assay vs. plating, 92
plate cultivation of thermophiles, 169
preparation and sterilization for large-scale fermentor growth of hyperthermophiles, 47
Pyrobaculum, 181
Pyrococcus/Staphylothermus, 180–181
Pyrodictium, 178–179
abyssi, 184
SF1, 179–180
simplified basal, for *Sulfolobus acidocaldarius*, 172–173
Stygiolobus, 185
Sulfolobus, 173
solfataricus, 175
Thermococcus
celer, 178
litoralis, 186
Thermofilum pendens, 177–178
Thermoplasma
acidophilum, 175
volcanium, 183
Thermoproteus, 176
neutrophilus, 179
TTD, 184
WMC, 182
Metallosphaera
commercially available strains, 160
sedula, growth using continuous culture techniques, 31–35
MetaPhor agarose, analysis of restriction digests using, 153
Methanococcus jannaschii, high-pressure–high-temperature cultivation, 37–40
Methanogenic Archaea
growth media and preparation
marine, 20
nonmarine, 21
properties, 5
Methanogenium growth medium, 173–174
Methanothermus, culture medium, 21
Methanothrix (thermophilic) medium, 181
Methyl viologen, as reduced redox dye, assay for hydrogenase, 67–68
MG-medium, 185
Mid-Atlantic Ridge, hydrothermal vent and isolation of heterotrophic hyperthermophiles, 9
Millipore tangential-flow filtration cell vs. continuous-flow centrifuge
contamination with SDS and cell lysis, 58
harvesting cells using, 56
Mineral sulfides, storage and transportation, 11
Modified accumulation method (A_0), measuring *Sulfolobus acidocaldarius* mutation rates, 133–135
Mono Q HR 5/5 column, hydrogenase/ferredoxin purification, 69–70
Monosaccharides, as carbon source for S^0-dependent Archaea growth, 4
MTP4 medium, 185
^{15}N-NMR spectrometry, study of glucose metabolic pathways in *Thermoplasma acidophilum*, 81–85
Norprene tubing, low oxygen permeability, 32
Novobiocin, spontaneous resistance mutant selection and, 127
Neutral lipid, total lipid fractionation and silicic acid column chromatography, 76–77
Neutrophiles, (nonmethanogenic) growth media and preparation, 17–19
Nitrogen (oxygen-free), preparation, 10
Nonidet P-40, 101, 103
Northern blot analysis of RNA, 99

ORFs (open reading frames), rRNA introns containing, 114
Oxygen toxicity
hyperthermophile transfer and, 11
and *Thermoplasma acidophilum* culture, 56

^{31}P-NMR spectrometry, 81
PAGE (polyacrylamide gel electrophoresis) two-dimensional, of total DNA, 108–109
Palladium cold catalyst, 25
PCR (polymerase chain reaction)
choosing forward and reverse primers, 105
isolation of 16S rRNA genes from hyperthermophilic Archaea, 101–105
subunit rRNA gene primers, 104
typing hyperthermophilic Archaea based on 16S/23S rRNA spacer regions and, 151–152
Peptone
amino acid composition, 12
as organic substrate in hyperthermophile culture medium, 10
pGT5, from *Pyrococcus abyssi*, 87
pH, *Thermoplasma acidophilum* culture contamination and, 56, 58
Photobiotin acetate, 119
Plaque assay
medium containing Gelrite, 19
transfection of *Sulfolobus solfataricus* and, 91–93
Plasmids, purification from thermophilic and hyperthermophilic Archaea
alkaline lysis method, 89
alternative protocol, 89
harvesting and lysis of the archaeon, 88
isolation of total DNA, 88–89
linearization, T7 RNA polymerase transcription, 113
purification of plasmid DNA, 89–90

Plastic garbage can production of *Thermoplasma acidophilum*, 51, 54–56
Plate cultivation technique for anaerobic, thermophilic, sulfur-metabolizing Archaea, 25–28
Polar lipid
fatty acids
ester-linked, 77
vs. whole-cell, 78
total lipid fractionation and silicic acid column chromatography, 76–77
Polarographic O_2 Sensor, 55
Primer
choosing forward and reverse primers for PCR, 105
deoxynucleotide synthesis, 116
extension, 99
forward and reverse, 149
oligodeoxynucleotide, 110–111
selection, PCR amplification of 16S rRNA genes and, 103–105
subunit rRNA gene primers, 104
Proteinase K, 88, 95, 149
pSL10, from *Desulfurolobus ambivalens*, 87
pSSV1, from *Sulfolobus shibatae*, 87, 91
pyrB, 135
pyrF mutants, 133
Pyrimidine auxotrophic mutants, direct selection from *Sulfolobus acidocaldarius*, 127–128
Pyrobaculum
commercially available strains, 160
medium, 181
Pyrococcus
abyssi
colonies, after incubation on Gelrite plate, 28
growth medium, 18
pGT5, 87
commercially available strains, 160–161
endeavori. See also ES4 cultivation
high-pressure–high-temperature cultivation, 37, 41–42
furiosus
basal salts medium, 169–170
continuous culture techniques and, 31–35
growth
on agar media, 18
in artificial sea water, 34
large-scale, 47–48
small-scale, 48
hydrogenase and ferredoxin purification, 67–71
growth medium, 18
/*Staphylothermus*, 180–181
strain GB-D, 12
growth on purified protein media, 18
Pyrodictium
abyssi, 64–65
medium, 184
commercially available strains, 161
medium, 178–179
spp., solubilization and purification of hydrogenases
abyssi, 64–65
brockii, 64
further purification, 65
Pyruvate, 85
ferredoxin oxidoreductase, 67

Q-Sepharose column chromatography, hydrogenase/ferredoxin separation, 69
Quinones, 63, 79

Rabbit reticulocyte lysate, detection of DNA cleavage activity, 116
Reactor systems. *See also* Fermentors
continuous culture of thermophilic and hyperthermophilic Archaea, 31–32
high-pressure–high-temperature, cultivation of thermophilic Archaea, 37–44
Replica-plating
incubation of plates, 131
isolation of auxotrophic mutants, 128–131
Resazurin, as redox indicator, 10, 13, 16, 41
solution
in methanogen growth medium, 20–21
in neutrophile growth medium, 17–18
Resistance mutants
selection of *Sulfolobus acidocaldarius* mutants and, 126–127
spontaneous, in strain DG6, 127
Reverse
mutation by P_0 method, measurement in *Sulfolobus acidocaldarius*, 135–137
primers, 149
choosing for PCR, 105
transcriptase, 109
RFLP (restriction fragment length polymorphism), 149
analysis of DNA polymerase genes, 142–147
typing marine vent thermophiles by DNA polymerase RFLP, 140
RNA. *See also* rRNA
extraction from sulfur-utilizing thermophilic Archaea, 97–99
linear and circular, mobility at various acrylamide concentrations, 108
mRNA, isolation from total RNA, difficulties, 123
rRNA
genes
of hyperthermophiles, archaeal introns, 107
proteins encoded by rRNA introns, 114–116
5S
gene sequences, 195–200
secondary structure, 200
16S
gene amplification by PCR, 101–105
-like primers, 201–203
spacer region, typing hyperthermophilic Archaea, 149–153
23S
rRNA-like primers, 201–203
spacer region, typing hyperthermophilic Archaea, 149–153
Rubredoxin, 67

S1 mapping of in vitro transcription, 157
Scanning calorimetry, setting up a scan
data analysis, 206–207
filling procedure, 206
instrument specifications for DASM-4, 107
sample preparation , 206
scanning, 206
SDS (sodium dodecyl sulfate), 103
plasmid lysis and, 87–89
Selenium, 20
Sephadex G
-50 column chromatography, T7 RNA polymerase transcription, 111
-100 column chromatography, preparation of ^{32}P-labeled S1 probe, 156–157
Sepharose CL-6 column chromatography, ferredoxin purification, 70
SF1 medium, 179–180
S^o
-dependent Archaea
as aerobic S^o-oxidizers, 4
as anaerobic S^o-reducers, 4
-reducing
cultures, small-scale vs. large-scale growth, 48
heterotrophs, 4
reduction to H_2S, 4, 48
Sodium
acetate, 111
PCR template purity and, 103
chloride
concentration in gel medium, 27–28
lipid phase separation and, 76
dithionite, as reducing agent, 10–11, 16
hydrogenase assay and, 67–68
separating hydrogenase and ferredoxin and, 69
hydroxide
scrubber, 42–43
scrubbing effluent gas from continuous culture reactor, 31

Sodium (*continued*)
sulfate, lipid phase separation and, 76
sulfide, addition to induce anaerobicity, 16
Solid
polymetal sulfides, hyperthermophile samples, 10
medium
containing elemental sulfur, 25–27
culture and growth of hyperthermophiles, 18
preparation using Gelrite, 19
Southern blot analysis
and DNA polymerase RFLP analysis of marine vent thermophiles, 142–145
of marine vent isolates, 146
of rRNA genes, 109
Springs and geysers vs. marine geothermal vents, effect of temperature on thermophiles, 6
Stainless steel
cooling coils for Sharples centrifuge, 47–48
fermentor, large-scale growth of hyperthermophiles, 47–48
high-pressure vessel, effect on cell growth/metabolism, 39–40
Staphylothermus
commercially available strains, 161
spp., growth medium, 18
Storage of heterotrophic hyperthermophiles, frozen vs. cooled samples, 11
Stygiolobus, 4
commercially available strains, 161
medium, 185
Submersible-coupled in situ sensing and sampling system (SIS^3), 16
Subtraction probes, generation for isolation of genes in thermophilic Archaea
creation of DNA from a population of differentially expressed genes, 121–122
hybridization of cDNA population to uninduced biotinylated RNA, 122–123
removal of ribosomal genes from total DNA, 120–121
Sulfate-reducing Archaea, properties, 5
Sulfides, solid polymetal, collection from deep-sea hydrothermal vents, 10
Sulfolobales, 59
properties, 5
as S^o-dependent Archaea, 4
Sulfolobus
acidocaldarius, 59
liquid culture aeration to attain high cell densities, 126
mutant isolation
auxotrophic mutants, isolation by replica-plating, 128–131
cultivation, 125–126
pyrimidine auxotrophs, direct selection, 127–128
resistance mutants, direct selection, 126–127
mutation rate measurements
forward mutation using A_0 method, 133–136
reverse mutation by P_0 method, 135–137
simplified basal medium, 172–173
brierleyi, commercially available strains, 159
commercially available strains, 161–162
growth medium, 19
liquid, 172
medium, 173
metallicus, 19
optimal growth conditions, 19
shibatae
in vitro transcription from natural and mutated rDNA promoters, 155–158
pSSV1, 87
solfataricus
medium, 175
transfection, 91–93
Sulfur
anaerobic growth on sulfur, selectivity for *Thermoplasma acidophilum*, 53
colloidal
addition to reactor system for continuous culture of thermophilic and hyperthermophilic Archaea, 32
in Gelrite plating medium for hyperthermophile culture, 19
precipitation in liquid culture, 19
preparation for continuous feed cultures, 33
streaking cell suspensions on agar plates containing, 11
crystallization on culture plate surface, 28
-dependent Archaea, properties, 5
elemental, 10, 17, 31, 59, 140
addition to *P. furiosus* growth medium and cell yield, 35
"Flowers of Sulfur," 52
replacement with cysteine in hyperthermophile culture medium, 10
solid medium containing, 25–27
oxidation to sulfuric acid, 4
-reducing hyperthermophiles. *See Pyrococcus furiosus*; *Thermococcus litoralis*
reduction to H_2S, 4
Superose 6 HR 16/50 column chromatography, hydrogenase purification, 70
T7 RNA polymerase, 112
in vitro splicing, 113
transcript, 111
transcription, 113
Thermal cycler
parameters, 103
programming example, 102
Thermoacidophiles, cultivation using continuous culture techniques, 35
Thermoanaerobes, cultivation using continuous culture techniques, 34–35
Thermococcus
celer
chromosome map of strain Vu13, 189–190
medium, 178
commercially available strains, 162
litoralis
DNA polymerase gene, 142
large-scale growth, 47–48
medium, 186
Thermofilum
commercially available strains, 162
pendens, medium, 177–178
Thermophilic Archaea (thermophiles). *See also* Hyperthermophilic Archaea
as aerobic S^o-oxidizers, 4
as anaerobic S^o-reducers, 4
classification, 3
as hyperthermophile, 4
codon usage tables, 191–194
continuous culture techniques, 31–36
discovery of deep-sea hydrothermal vents and, 3
first established as life form, 3
high-pressure–high-temperature cultivation
gas and liquid sampling, pressures
≥50 atm, 42–43
<50 atm, 43
Methanococcus jannaschii, 37–40
Pyrococcus endeavori (ES4), 40–42
isolation of thermophilic microorganisms, 3
marine vent, typing by DNA PCR RFLPs
chromosomal DNA isolation, 140–142
growth of marine vents, 139–140
RFLP analysis, 142–145
optimal growth conditions, 3, 6
properties, 5
subtraction probe generation for isolation of specific genes, 119–123
sulfur
-dependent, 4
-utilizing, RNA extraction, 97–99
thermodynamic data on activation and denaturation, 205–207
typing based on 16S/23S rRNA spacer region, 149–153
Thermococcales, 4
properties, 5

Thermoplasma acidophilum, culture
aerobic culture, 82
boiling medium before inoculation, 52
cell morphology, 52
eliminating contaminating organisms, 52
small-scale, 51–52
on sulfur, 52–54
transferring cells and decline of cell densities, 52
cleaning and storing filtration cell, 57
distinguishing cells from sulfur particles, 54
glucose metabolic pathways, study using ^{13}C- and ^{15}N-NMR spectrometry, 81–85
growth medium, 171–172
H_2S production, 53
large-scale aerobic cell growth
harvesting cells, 56–57
alternative harvesting procedure, 57–58
culture medium, 55–56
inoculum, 55
lysis, 82
maintaining established cultures, 54
medium, 175
buffering capacity, 57
NMR spectrometry of cell lysates, 82
O_2 toxicity, 56
partial neutralization, 57
pH and culture contamination, 56, 58
plastic garbage can production, 51, 54–56
preparing cell lysate, 60–61
spp., commercially available strains, 163
storage of cultures, 59–60
sucrose vs. glucose, 59
yeast extract, addition to large-scale culture, 56
yield, 58
Thermoplasma volcanium medium, 183
Thermoplasmatales, properties, 5
Thermoproteales, 4
properties, 5
Thermoproteus
commercially available strains, 163
medium, 176
neutrophilus, 179
Titanium
citrate, as reducing agent, 20, 34
syringe, obtaining hot water hyperthermophile samples using, 15
Torbal jar, incubating streaked hyperthermophile suspensions on agar plates, 11
Trace elements A and B, in neutrophile media, 17
Transcription from natural and mutated rDNA promoters, 155–158
Transfection of *Sulfolobus solfataricus* with SSV1 DNA
efficiency/frequency using DNA uptake method, 91
electroporation, 92
residual salts, effect on time constant during electroporation, 92
Transportation of heterotrophic hyperthermophiles, optimal temperature, 11
Trypticase soy, 17
Tryptone, 48
cultivation of thermoanaerobes and, 34
as nitrogen source in isolation of amino acid auxotrophs, 130
selective media for P_0 method and reverse mutation, 136
TTD medium, 184
Tungsten, growth of hyperthermophiles and, 22, 48
Two-dimensional
heteronuclear multiple quantum coherence (HMQC), 83, 85
PAGE, total DNA analysis, 108–109

Vitamin mixture, 12

WMC medium, 182

Xylose-tryptone plates, pyrimidine auxotroph direct selection and, 127

Yeast extract
addition to large-scale culture of *Thermoplasma acidophilum*, 56
optimal concentration, 59
amino acid composition, 12
cultivation of thermoanaerobes and, 34
as organic substrate in hyperthermophile culture medium, 10, 48
substituting brands, small-scale growth of *Thermoplasma acidophilum*, 52